Strategies for the
Control of Cereal Disease

Frontispiece   Lodging of winter wheat (cv. Squarehead's Master)
caused by eyespot (*Pseudocercosporella herpotrichoides*) on Broadbalk
Field, Rothamsted in 1943. Such devastation due to eyespot can now
be avoided by using short-strawed, resistant cultivars and systemic
fungicides. Photograph Rothamsted Experimental Station.

Strategies for the Control of Cereal Disease

Edited for the
Federation of British Plant Pathologists
by
J. F. JENKYN and R. T. PLUMB

Rothamsted Experimental Station,
Harpenden, Hertfordshire

BLACKWELL SCIENTIFIC PUBLICATIONS
OXFORD LONDON EDINBURGH
BOSTON MELBOURNE

ⓒ 1981 Blackwell Scientific Publications
Osney Mead, Oxford OX2 OEL
8 John Street, London WC1N 2ES
9 Forrest Road, Edinburgh EH1 2QH
52 Beacon Street, Boston, Massachusetts 02108, USA
214 Berkeley Street, Carlton, Victoria 3053, Australia

First published 1981

British Library
Cataloguing in Publication Data

 Strategies for the control of cereal disease
    1. Grain - diseases and pests
    I. Jenkyn, J.F. II. Plumb,R.T.
    III. Federation of British Plant Pathologists
    633.1'0493   SB608.G6

 ISBN 0-632-00716-8

Distributed in the USA and Canada
by Halsted Press, a division of
John Wiley & Sons Inc, New York

Printed in Great Britain

The Federation of British Plant Pathologists is part of the Association of Applied Biologists and the British Mycological Society

Contents

Section 1
Host Resistance

Preface

In 1979 the Association of Applied Biologists celebrated its 75th
Anniversary by organizing a major international meeting with the title
"Advances in Crop Production and Crop Protection", held at the
University of Reading. As their contribution to this meeting the
Federation of British Plant Pathologists organized a programme to
consider "Strategies for the Control of Cereal Disease". This book is
based on the papers presented at that meeting but is not intended to
constitute a report of the proceedings.

The three sections into which the volume is divided deal with host
resistance, chemical control and husbandry, and are based on the corresponding sessions of the programme. Each is briefly introduced by the
chairman of that session.

Based as it is on a scientific meeting, the volume is inevitably
selective rather than comprehensive. It has, for example, not been
possible to cover the tropical cereals. Nevertheless, the diverse
interests and experience of the authors provide a balanced coverage of
the present state of disease control in temperate small-grain cereals
and of some likely developments in the future. Because cereals are so
important throughout the world, the book should contain something of
interest to all those involved with their cultivation. However, some
of the approaches described have wider application and we believe the
volume contains much that has relevance to those working with other
crops.

On behalf of the Federation we thank the authors for their co-
operation in the preparation of the volume. We also thank Eileen M.
Ennever and Fran Cheesman who prepared the camera-ready typescript and
Margaret J. Howe who compiled the index. Thanks are also due to
B.D.L. Fitt, who was the production editor. Finally we are very
grateful to P.R. Scott who not only organized the meeting but, as
F.B.P.P. Publications Representative gave us invaluable guidance and
support throughout the preparation of this volume.

J.F. JENKYN
R.T. PLUMB

Federation of British Plant Pathologists

Section 1
Host Resistance

Chairman's comments

J. K. DOODSON

National Institute of Agricultural Botany,
Huntingdon Road, Cambridge

Resistant cultivars have been the mainstay of cereal disease control for many years and they continue to be an essential component of integrated disease control methods. Ideally, if single genes conferring effective and long-lasting resistance to each of the important diseases were available, diseases would cease to have such an important influence on both breeders and farmers. The farmer would have to worry less about the effects of husbandry on diseases and would obviously be freed of the expense and inconvenience of using other methods of disease control. The breeder would be able to concentrate his efforts in other directions, especially improving yield and grain quality.

Unfortunately we are far from achieving this ideal. Against some diseases (e.g. take-all) little useful resistance has yet been identified and with others (e.g. barley yellow dwarf virus) it seems difficult to incorporate the known resistance genes into acceptable commercial cultivars. Genetic resistances have been successful in decreasing the level of some diseases (e.g. eyespot in wheat) but while resistance genes used against the air-borne, biotrophic pathogens, such as the rusts and mildews, can be extremely effective, their useful life is often short because virulences able to overcome them may be rapidly selected in the pathogen population. As a result we have had to look for alternative, long-lasting resistances and to consider new strategies for exploiting cultivars with single gene resistances, so that their usefulness is prolonged.

The old adage not to put all one's eggs in one basket is particularly apt when choosing cultivars and farmers are well advised to grow a selection of cultivars with different resistances. The recommended lists of cereal cultivars produced by the NIAB now incorporate specific advice to help farmers diversify the cultivars they grow (Priestley). Such diversification between fields should decrease the risk of severe epidemics occurring over a large area but ideally we should also aim to increase diversity within fields. Multiline cultivars (Browning & Frey) have the advantage that they are heterogenous for resistance to specific diseases while being isogenic for other characters. However, much breeding effort is required to produce a multiline effective against just one disease let alone the several which normally occur.

Mixtures of existing cultivars with different resistances seem to be very effective in reducing mildew levels (Wolfe *et al*.) and have the advantage that the components of the mixture can be easily changed to take advantage of new sources of resistance. The choice of components will obviously be affected by their resistance to diseases other than mildew and their agronomic characters. The limited evidence available (Jeger *et al*.) suggests that cultivar mixtures will also be useful in controlling non-specialized and splash-dispersed pathogens.

Although specific genes are relatively easy to manipulate some breeders have come to accept that to control many of the air-borne diseases, more durable forms of resistance are essential. There is, therefore, much interest in identifying and understanding such resist-ances so that they may be exploited most efficiently (Parlevliet, Johnson).

The most economical way to control diseases is by growing resistant cultivars and the incorporation of resistance into high yielding wheat cultivars is a good illustration (Bingham). However, it is appropriate to be reminded by the breeders that they have to consider many other characters and, as pointed out by Habgood & Clifford, the end product represents a compromise between the desirable and the possible. The problems faced by those breeding cultivars for international use are even more complex as they must produce material adapted to a wide range of climates and incorporate resistances to as many diseases as the cultivar is likely to encounter (Dubin & Rajaram).

The importance of cereal diseases is now widely recognized and strategies to control them will increasingly integrate all available methods. However, it is clear from the papers which follow, that plant breeding will continue to play an important role in the foreseeable future.

Breeding wheat for disease resistance

J. BINGHAM
Plant Breeding Institute,
Trumpington, Cambridge

OBJECTIVES IN BREEDING FOR DISEASE RESISTANCE

A substantial proportion of the fundamental and strategic research in
plant pathology and other areas of botanical investigation has its
ultimate application in the breeding of new cultivars. However, there
are limitations on the sizes of plant populations which can be handled
efficiently so it is not practicable for breeders to work intensively
with every character suggested by their colleagues. It follows that a
well informed and astute choice of objectives is the key to an effective
plant breeding programme. The breeding effort must not be dissipated by
objectives which are of little value in practice or are manageable
economically by husbandry. It is equally true that many cultivars have
failed because the breeder overlooked a single character which assumed
importance during the course of a breeding programme.

Rapid progress in the development of fungicides for use on cereals
has led some chemical manufacturers, agronomists and farmers to consider
disease resistance as a relatively unimportant character in determining
their choice of cultivars. This has been taken to the extreme in
Schleswig-Holstein (Effland, this volume) where the principal wheat
cultivar, Vuka, is grown because it has good bread-making quality and
yield potential, despite being very susceptible to yellow rust (*Puccinia
striiformis*), mildew (*Erysiphe graminis*), and eyespot (*Pseudocercosporella herpotrichoides*), and also having long, weak straw. The
official trials in that area of the Federal Republic of Germany are com-
prehensively treated with fungicides and growth regulators; no account
is taken of disease resistance when recommending cultivars so breeders
are in a quandary when setting priorities for disease resistance.

In the United Kingdom (UK) a minority of farmers on the most productive soils, where wheat regularly yields 8 t/ha or more, apply
fungicides prophylactically, typically on two occasions to control
eyespot and foliar diseases. It is likely that such productive crops
will continue to justify the cost of this policy, though experimental
work has shown that greater profits can often be obtained by relating
fungicide use more closely to cultivar resistance and observations of

disease occurrence (Anon, 1980a). For average crops, where yield is limited by soil type and moisture availability, costs are expected to become critical and the trend is towards a single application of a broad-spectrum fungicide at ear emergence to give protection during grain filling.

Genetic improvements in the yield potential of new wheat cultivars, measured in the absence of disease and lodging, are currently averaging not much less than 1% per annum, and it seems probable that this rate of progress can be maintained for at least 20 to 30 years (Austin *et al.*, 1980). Assuming this is feasible the most appropriate allocation of resources is in breeding cultivars that need no more than a single fungicide application at ear emergence. Resistance would then have two functions; firstly to give protection throughout the vegetative phase so that sprays before anthesis are unnecessary and secondly to provide, in combination with a fungicide, adequate control after anthesis.

With this strategy more emphasis should be given to eyespot than at present because resistance derived from cv. Cappelle-Desprez, and now incorporated in most cultivars grown in the UK, does not necessarily obviate the need for an early fungicide treatment. Moreover, eyespot is a wheat disease which is difficult for the farmer to monitor and is likely to become more important now that the area of winter cereals in the UK has increased. It would also be desirable to give more attention to resistance to brown rust (*Puccinia recondita*) which has become more important due to the cultivation of susceptible cultivars and the greater use of nitrogen fertilizers. A particular problem with this late-developing disease is that fungicides applied at ear emergence, which is often before infection is evident, may not give adequate control.

The term "broad-spectrum", usually used to describe fungicides, is equally relevant to cultivar resistance, and of great importance since a weakness in resistance to any one disease may necessitate fungicide treatment and largely negate the value of resistance to other diseases. To avoid this predicament cultivar resistance should be not less than that regarded as "desirable" by the National Institute of Agricultural Botany (Anon, 1980b) and should probably be greater for eyespot and mildew.

BREEDING PROCEDURES

Except for a very small area of *Triticum durum*, the wheat cultivars grown in the UK belong to the bread wheat species, *T. aestivum*. The cultivars are naturally self-pollinating and those accepted for the National List and grown commercially are highly inbred and true-breeding. The genetic variation needed for selection purposes is mostly obtained by the hand-crossing of cultivars or breeding lines. The following generations are then allowed to self-pollinate. Genetic segregation generates diverse populations in F_2 and with each succeeding generation the individual plants become progressively more homozygous. A uniform cultivar is typically based on a single F_5 or F_6 plant.

Much of the practice of breeding is aimed at detecting the rare plants which have an improved combination of characters. Various systems of selection have been devised, including the pedigree trial method used at the Plant Breeding Institute (Bingham, 1979). In the winter wheat project about 600 crosses are made each year and 1500 plants of each cross grown as spaced plants in the field in F_2. Following visual selection for agronomic type and disease resistance about 35 000 of these plants are retained and sown as separate ear-row observation plots in F_3. This process of ear to row selection is repeated for several generations until the lines are stable. Yield trials begin at one site in F_5, and in the year prior to National List Trials about 20 lines are still under consideration in yield trials at nine centres. The most promising lines are entered for National List Trials in F_9 and the first distribution of basic seed by the National Seed Development Organisation is normally in F_{12}.

The main departure from this system is in the use of partial or complete backcrossing to introduce features which are fairly simply inherited and readily identified. Cv. Maris Huntsman was selected from a crossing programme of this type, with one backcross to cv. Professeur Marchal to re-establish a good agronomic type with resistance to mildew derived originally from *T. timopheevi* via the Wisconsin line CI 12633 (Fig. 1). As only one backcross was made there was sufficient residual variation to allow selection for other characters, including potential yield and straw strength. When transferring polygenically inherited characters from outdated cultivars, or from alien agronomic types poorly adapted to the local environment, it is common practice to select the original cross intensively for one or two generations before making the backcross.

Mutation breeding is now rarely used by wheat breeders and there are probably no commercial cultivars in north Europe with significant, new genetic factors for disease resistance which have been produced by this method. Ionizing radiations have, however, been used to increase genetic recombination, especially in the transfer of disease resistance from alien species (Sears, 1956). Interest in F_1 hybrid cultivars has also declined, mainly due to difficulties in the commercial production of seed using cytoplasmic forms of male sterility. Investigations are, however, continuing into the development of chemical male sterilants, and a cytogenetic system has been devised which may make it possible to use a nuclear gene for male sterility (Driscoll, 1972). The development of an economic method for producing F_1 hybrid cultivars would offer a useful new approach to breeding for disease resistance, especially in the possibility of exploiting allelic variation, and allowing greater speed and flexibility in the combination of non-allelic dominant factors.

Cytoplasmic or nuclear factors for male sterility are also being used by a number of breeders to establish continuously cross-breeding

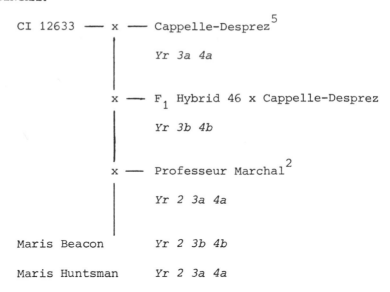

Figure 1. Pedigree of cv. Maris Beacon and cv. Maris Huntsman showing the genes for seedling reaction to yellow rust carried by each cultivar.

populations for selection purposes (Laabassi, 1979). Any desired degree of cross pollination can be maintained from generation to generation by varying the proportion of male-sterile plants in the population. After several cycles of selection, normally fertile lines are taken from the population and entered into a conventional pedigree system. The attraction of this breeding method is in the opportunity it provides for assembling a number of minor genes for disease resistance which might not be individually detectable. In wheat it is being used by Scott & Hollins (1976) to investigate the possibility of obtaining a useful level of resistance to take-all (*Gaeumannomyces graminis*). This method would be difficult to apply to diseases such as yellow rust where there are many factors for specific resistance which in practice cannot be excluded from the population. With this method the breeder forfeits the degree of control over his material which is inherent in the pedigree system; if a line with unrecognized defects is included in the population, undesirable genes are rapidly dispersed and difficult to eliminate, whereas with the pedigree system the offending cross is discarded in the normal course of selection.

Thus it seems certain that the main breeding effort will continue to follow the pedigree method first used by Biffen in the early years of this century, and breeders will continue to seek tests which will improve the efficiency of selection. As the first stage in selection it is generally considered essential to establish artificial epiphytotics of all the relevant diseases throughout the ear-row material. At the Plant Breeding Institute, susceptible spreader cultivars are infected with about ten races of yellow rust and two of brown rust. The ear-rows are also sprayed with an inoculum of *Septoria nodorum*. Infected stubble is used to spread eyespot, and mildew is encouraged by the use of susceptible spreader cultivars and high rates of nitrogen fertilizer.

CASE HISTORIES OF THE PRINCIPAL DISEASES OF WHEAT

Yellow rust may be considered in detail as a good example of the problems encountered in breeding for resistance to those foliar diseases which have the capability for evolving new physiologic races.

Over the last 15 years more than 10 new, or previously unidentified, races of yellow rust have been found in the United Kingdom. Many of the cultivars and breeders' lines affected by these new races had previously shown hypersensitive resistance to existing races in seedling tests. There is, however, good evidence that durable resistance to yellow rust does exist (Johnson, 1978), in that resistance remained effective over long periods of cultivation in some cultivars, notably cvs Browick, Little Joss, Yeoman, Vilmorin 27, Atle, Hybride de Bersée, Cappelle-Desprez and Maris Widgeon. These cultivars are all susceptible at the seedling stage but have sufficient resistance in the adult plant to restrict the development of epidemics. However, selection for seedling susceptibility and adult plant resistance does not in itself guarantee durability of such resistance at the original level, as is evident from the discovery of isolates adapted to infect adult plants of cv. Joss Cambier in 1971, cvs Maris Bilbo and Hobbit in 1973 and cv. Maris Huntsman in 1974. The breeding problem is to be able to recognize and handle the different categories of resistance in large numbers of single-plant progenies.

Wheat cultivars commonly have genetic factors for seedling resistance to yellow rust, including the genes $Yr 1$ to $Yr 7$ (Table 1) identified by Lupton & Macer (1962), Macer (1966) and others. It is probable that many more such factors could be found by searching collections of cultivars and land races. Genes of similar effect have also been introduced from related species, for example $Yr 8$ in cv. Compair as a chromosome 2M/2D translocation from $Aegilops$ $comosa$ (Riley, Chapman & Johnson, 1968), and $Yr 9$ in the cvs Riebesal 47/51, Kavkaz and Clement as a 1R/1B substitution or translocation from rye. However, the durability of such resistance in agricultural use is unpredictable and all the gene combinations possessed by the cultivars in Table 1 have been overcome.

Moreover, in our experience, multiple factor seedling resistance (i.e. with more than one gene or combinations of genes each giving seedling resistance to all known races) has not been usefully long lasting. The genes incorporated in cv. Maris Beacon from cvs Professeur Marchal and Hybrid 46 (Fig. 1) were soon overcome separately and then in combination. Cv. Hybrid 46 was found to be susceptible to race 3/55 (40 E9) in 1967, cv. Professeur Marchal to race 58C (41 E136) in 1968 and cv. Maris Beacon to the combined virulence race 3/55D (104 E137) in 1969; none of these cultivars had any underlying adult plant resistance. Cv. Maris Huntsman, from the same cross, was a breeding failure in the

Table 1. Genes for seedling reaction to yellow rust in some wheat cultivars.

Cultivar	*Yr* genes									
	1	2	3a	3b	4a	4b	6	7	8	9
Cappelle-Desprez			+		+					
Hybrid 46				+		+				
Maris Beacon		+		+		+				
Maris Huntsman		+	+		+					
Hobbit			+		+					
Bounty	+		+		+					
Kinsman			+		+		+			
Talent								+		
Compair									+	
Clement										+

sense that the objective of combining seedling resistance factors was not achieved. However, it was selected for adult plant resistance which proved more durable and, even since the emergence of the more virulent form of race 41 E136 in 1974, it has retained a level of resistance which is adequate in most situations.

Many breeders have had similar experiences and it is now common practice at the Plant Breeding Institute to avoid lines which are seedling-resistant to all known races. This policy entails extensive seedling tests because new combinations of factors giving such resistance occur frequently in breeding programmes. For example crosses of the French cultivar Talent and the Dutch cultivar Clement with cultivars from the Plant Breeding Institute, including cvs Maris Huntsman, Hobbit, Virtue, Hustler, Bounty, Avalon and Brigand (Table 1), segregate a high proportion of seedling-resistant lines. Such crosses are considered necessary to obtain genetic diversity in yield.

Although it is clear that durable forms of adult plant resistance to yellow rust do exist, selection is greatly hampered by the lack of a test for durability. Crosses and selection methods can, however, be planned to increase the probability of obtaining durable resistance (Johnson, 1978; this volume). For example the winter wheat cultivar Bounty was selected from a cross of cv. Ploughman with cv. Durin (Fig. 2), (Bingham *et al.*,1980). Cv. Ploughman resulted from a partial backcross programme to cv. Maris Widgeon and is therefore assumed to have durable adult plant resistance from this source. Cv. Ploughman also has genes for seedling resistance to race 41 E136, derived from cv. Hybrid 46, which are not present in cv. Bounty. Cv. Durin is susceptible to race 41 E136 both as a seedling and as an adult plant. The adult plant resistance of cv. Bounty was selected using race

41 E136, so it is probably similar to that of cvs Ploughman and Maris
Widgeon and may be expected to be durable. A similar procedure was
followed in the breeding of cv. Avalon.

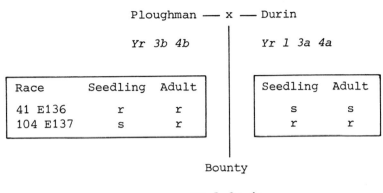

Figure 2. Pedigree of cv. Bounty showing the genes for seedling
resistance to yellow rust, and the reaction of cv. Bounty and its
parents to two races of yellow rust in the seedling and adult plant
stages (r, resistant; s, susceptible).

By contrast, the breeding of cv. Brigand, from a cross of cv. Maris
Huntsman with cv. Bilbo, illustrates how adult plant resistance assemb-
led from more susceptible parents (Johnson, 1978) may not be durable.
Cvs Maris Huntsman and Bilbo are both susceptible as seedlings to the
races 41 E136 and 104 E137 but differ in their adult plant reactions.
Adult plants of cv. Maris Huntsman are moderately susceptible to race
41 E136 type 3 (originally isolated from cv. Maris Huntsman) but resistant to 104 E137 type 3 (originally isolated from cv. Bilbo). Conversely adult plants of cv. Bilbo are resistant to 41 E136 type 3 but
very susceptible to 104 E137 type 3. Until 1979 cv. Brigand had shown
a high level of adult plant resistance to both races. However, it seems
probable that this resistance was dependent on a race-specific component
from cv. Bilbo because a race has since been found (Anon, 1980b) to
which cvs Brigand and Maris Huntsman are moderately susceptible.

The ideal strategy to control yellow rust might be to superimpose
genetic factors for seedling resistance on a background of durable adult
plant resistance. Attempts to do this by backcrossing to adult plant
resistant lines have made little impact in the UK because the recurrent
parents have usually been superseded during the course of the work by
other cultivars with improved yield or grain quality. An alternative

method, based on the pedigree system, is to select families which are
segregating a gene for seedling resistance and test these for adult
plant reaction. If the seedling-susceptible lines of such families are
uniform for adult plant resistance, it may be inferred that the lines
which are homozygous for seedling resistance have the same background of
adult plant resistance.

Similar considerations apply to breeding for resistance to mildew.
Genetic factors for seedling resistance were involved in the breeding of
cvs Kolibri, Maris Huntsman, Maris Dove, Armada, Flanders and Timmo.
New races which are able to overcome the genes concerned have since
occurred, though some of these cultivars have continued to exhibit some
adult plant resistance. Durable forms of adult plant resistance to
mildew do exist, for example in cvs Cappelle-Desprez, Maris Widgeon,
Maris Huntsman, Flinor and Bouquet. It is, however, less effective than
the adult plant resistance used against yellow rust. Thus there is a
need for new sources of resistance against mildew.

Experience outside the UK has shown that new races of brown rust can
also readily evolve. In the relatively unfavourable climate of the UK
it should be well within the power of the breeder to obtain good control
by resistant cultivars. The frequency of this disease in recent years
has demonstrated the capability of a minor disease to become important
when susceptible cultivars are grown, and emphasizes the need to select
for resistance.

Whereas obligate parasites of cereals tend to show host-specificity
equally to species and to cultivars, facultative parasites show specificity towards different species much more than towards different
cultivars (Scott et al., 1980). Thus with the facultative parasites
in the UK (notably Septoria nodorum, eyespot and take-all) the principal
breeding problem is not in the durability of resistance but in finding
sources of resistance which are sufficiently effective.

There are useful sources of resistance to Septoria nodorum in cvs
Hybrid 46, Champlein, Maris Huntsman and a number of exotic lines. Many
of the early semidwarf cultivars were rather susceptible   Their dense
canopy favours the spread of this pathogen, but other factors are
probably also involved in this tendency for them to be susceptible
(Scott & Benedikz, 1978). Fortunately, resistance from sources such as
cv. Maris Huntsman can still be effective in semidwarf cultivars,
although it is more difficult to select for resistance in these than in
taller cultivars.

Useful resistance to eyespot was first discovered in cv. Cappelle-
Desprez and was one of the most important characteristics of that
cultivar, contributing towards its consistency of performance and
explaining its use as a parent of many other cultivars. The relatively
simple inheritance of this resistance (Law et al., 1975) makes it easy
to select in breeding material, and all of the current winter wheat
cultivars bred at the Plant Breeding Institute have this resistance.
Surprisingly, it is only in the UK that resistance to eyespot has been

regularly sought by plant breeders and by the testing authorities, and
the possession of resistance equivalent to that of cv. Cappelle-Desprez
by 13 of the 16 cultivars of winter wheat recommended by the National
Institute of Agricultural Botany (Anon, 1980b) is unique.   There is now
an excellent prospect of further advance based on the line VPM 1 which
has resistance derived from *Aegilops ventricosa* (Doussinault *et al*.,
1974).   The resistance of VPM 1 to eyespot is greater than that of cv.
Cappelle-Desprez (Fig. 3) and genetic studies have indicated that it is
additive to that of cv. Cappelle-Desprez.   Combination of these resistances may obviate the need for fungicides even in short rotations.
However, there is some doubt whether the resistance from *Ae. ventricosa*
will be as durable as that from cv. Cappelle-Desprez.   There are
variants of the pathogen adapted to infect species other than wheat,
including *Ae. squarrosa*, although none adapted to *Ae. ventricosa* have
yet been found (Scott & Hollins, 1980a).

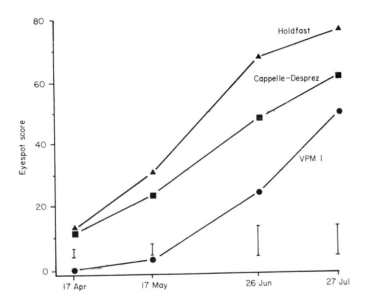

Figure 3.   Development of eyespot on three winter wheat cultivars in
a field trial at the Plant Breeding Institute in 1973.   This was
scored on a 0-100 scale, 100 indicating all stems severely infected.
Bars indicate LSDs, $P = 0.05$.

There is little prospect, at least within the next 10 years, of
achieving a substantial improvement in the resistance in wheat cultivars
to take-all because there are no sources of resistance which can be
readily utilized (Scott, 1981).   Although readily visible symptoms of
take-all are decreased by nitrogen fertilizers it is likely that resistance would increase the efficiency with which nitrogen is used by crops
and should therefore be actively sought.   Recent investigations have
shown that the resistance of rye is significantly expressed in most

triticale selections (Table 2). However, the genetic control of the rye
resistance may be too complex for it to be readily transferred to wheat.
As already mentioned it may also be possible to accumulate minor genetic
resistance factors from wheat, each of which has too small an effect to
be detected, by recurrent selection in an interbreeding population
derived from cultivars of diverse origin (Scott & Hollins, 1976).
Although similar events during the evolution of wheat have not produced
detectable variation in resistance, this could be because take-all was
seldom a selective influence prior to the cultivation of pure stands of
wheat in close rotations.

Table 2. Take-all score for wheat, rye and triticale in inoculated
field plots. (Scored on a 0-100 scale related to the proportion of
infected roots.)

	1977	1978	1979
Wheat (5 cultivars)	8.3	9.7	35.6
Rye (3 cultivars)	1.9	1.5	6.2
Triticale (4 lines)	4.0	4.6	17.2
SED	2.4	1.5	7.9

Data from trials described by Scott & Hollins (1980b).

FUTURE BREEDING OBJECTIVES

In the long term it must be expected that genetic improvements in yield
potential will become smaller and more difficult to obtain. As the rate
of improvement in yield declines there will be changes in emphasis on
other objectives; breeders will apply more resources to characters which
are not at present essential for the agricultural success of a cultivar.
For example, it may be possible to define grain quality more precisely
using nutritional criteria and there may be opportunities for specific
adaptation of cultivars to soil types or farming systems. In studying
disease resistance fundamental investigations should be strongly encouraged, with the ultimate objective of enabling the breeder to produce
cultivars which have very good, durable resistance not requiring support
by fungicides. However, as at present, very good resistance against
only one disease will be of little practical value and it will be essential to seek cultivars with broad resistance to all important pathogens.

ACKNOWLEDGEMENTS

I am grateful to all my colleagues at the Plant Breeding Institute who
have collaborated in the wheat breeding programme, and especially to
P.R. Scott and R. Johnson for their help in preparing this paper.

REFERENCES

Anon. (1980a) The use of fungicides and insecticides on cereals 1980.
    (Booklet 2257), *Ministry of Agriculture, Fisheries and Food
    (Publications)*, Pinner.
Anon. (1980b) *Recommended varieties of cereals*. National Institute of
    Agricultural Botany, Cambridge.
Austin R.B., Bingham J., Blackwell R.D., Evans L.T., Ford M.A., Morgan
    C.L. & Taylor M.T. (1980) Genetic improvements in wheat yields since
    1900 and associated physiological changes. *Journal of Agricultural
    Science, Cambridge* 94, 675-89.
Bingham J. (1979) Wheat breeding objectives and prospects.
    Agricultural Progress 17, 1-17.
Bingham J., Blackman J.A. & Angus W.J. (1980) Wheat, a guide to
    varieties from the Plant Breeding Institute. *National Seed
    Development Organisation*, Newton, Cambridge. 40 pp.
Doussinault G., Koller J., Touvin H. & Dosba F. (1974) Utilisation des
    géniteurs VPM 1 dans l'amélioration de l'état sanitaire du blé
    tendre. *Annales de l'Amélioration des Plantes* 24, 215-41.
Driscoll C.J. (1972) XYZ system of producing hybrid wheat. *Crop
    Science* 12, 516-7.
Johnson R. (1978) Practical breeding for durable resistance to rust
    diseases in self-pollinating cereals. *Euphytica* 27, 529-40.
Laabassi M.A. (1979) Genetic male sterility in wheat: cytogenetic
    analysis and application to breeding procedures. Ph.D. thesis,
    University of Cambridge.
Law C.N., Scott P.R., Worland A.J. & Hollins T.W. (1975) The inherit-
    ance of resistance to eyespot (*Cercosporella herpotrichoides*) in
    wheat. *Genetical Research, Cambridge* 25, 73-9.
Lupton F.G.H. & Macer R.C.F. (1962) Inheritance of resistance to
    yellow rust (*Puccinia glumarum* Erikss. & Henn.) in seven varieties
    of wheat. *Transactions of the British Mycological Society* 45, 21-45.
Macer R.C.F. (1966) The formal and monosomic genetic analysis of
    stripe rust (*Puccinia striiformis*) resistance. Proceedings of the
    2nd International Wheat Genetics Symposium (Ed. by J. MacKey)
    Hereditas Supplement 2, 127-42.
Riley R., Chapman V. & Johnson R. (1968) The incorporation of alien
    disease resistance in wheat by genetic interference with the
    regulation of meiotic chromosome synopsis. *Genetical Research,
    Cambridge* 12, 199-219.
Sears E.R. (1956) The transfer of leaf-rust resistance from *Aegilops
    umbellulata* to wheat. *'Genetics in Plant Breeding'*, Brookhaven
    Symposia in Biology 9, 1-21.
Scott P.R. (1981) Variation in host susceptibility. *Biology and
    Control of Take-all* (Ed. by M.J.C. Asher & P.J. Shipton).

Academic Press, London.   (In Press.)

Scott P.R. & Benedikz P.W. (1978)   Septoria.   *Report of the Plant Breeding Institute for 1977*, pp. 128-9.

Scott P.R. & Hollins T.W. (1976)   Take-all.   *Report of the Plant Breeding Institute for 1975*, pp. 134-5.

Scott P.R. & Hollins T.W. (1980a)   Pathogenic variation in *Pseudocercosporella herpotrichoides*.   *Annals of Applied Biology* 94, 297-300.

Scott P.R. & Hollins T.W. (1980b)   Take-all.   *Report of the Plant Breeding Institute for 1979*, p. 87.

Scott P.R., Johnson R., Wolfe M.S., Lowe H.J.B. & Bennett F.G.A. (1980) Host-specificity in cereal parasites in relation to their control. *Applied Biology* 5, 349-93.

Breeding barley for disease resistance: the essence of compromise

R. M. HABGOOD & B. C. CLIFFORD
Welsh Plant Breeding Station, Aberystwyth, Dyfed

RESISTANCE BREEDING POLICY

Most pathologists agree that the ideal resistance breeding strategy for barley, as for other field crops, would be to use "horizontal" resistance to control the important air-borne foliar pathogens and, where necessary, "vertical" resistance to control soil-borne pathogens. However, many pathologists also recognize that this objective is not easily achieved, and several alternative strategies, based on more or less sophisticated schemes of "gene management" (i.e. the spatial or temporal deployment of vertical resistance) have been proposed. These aim to prolong the effective life of resistance in commercial cultivars, which has frequently been brief when simple vertical resistance was used alone. Unfortunately, although these strategies may be in the national interest, most of them are of limited value to the individual breeder who is operating a commercial programme in competition against all other breeders. For the commercial breeder, resistance breeding is always a compromise between what is desirable and what is economically possible.

In any commercial breeding programme, the fundamental consideration in formulating policy is the necessity to produce cultivars which farmers will choose to grow, so that further breeding can be sustained by royalties levied on the sale of seed of these cultivars. Failure to achieve this objective must result either in the termination of the programme, or in its maintenance by public or private subsidy in spite of its minimal agricultural impact. The intensity of competition between barley breeders is illustrated by the fact that in the last 4 years, over 40 breeders from nine countries have submitted 170 winter and spring barley cultivars for United Kingdom (UK) National List testing, but only 28 cultivars, representing the products of 14 breeders in six countries, are included in the 1979 Recommended Lists of the National Institute of Agricultural Botany (NIAB) and of the Scottish Colleges.

Under these circumstances, no programme can hope for success unless its resources are concentrated exclusively on the immediate objective of producing cultivars which can be successfully marketed. Dissipation of

Jenkyn J. F. & Plumb R. T. (1981) *Strategies for the Control of Cereal Disease*

effort into avenues which are either speculative or superfluous imposes a crippling handicap on the programme, so it is essential to review the importance of disease resistance relative to the other characters in the programme. The value of disease resistance to a commercial breeder is represented by the additional seed sales which it will generate, rather than its value as a national asset, and must be compared with the market value of the improvements which could have been made in other directions if the resources had not been used for resistance breeding. Breeding for resistance inevitably incurs some penalty in yield potential, not necessarily through linkage of resistance genes with low yield (although this is known), but through the restricted choice of parents and progeny available for crossing and selection.

Analysis of the market value of disease resistance must recognize that there is not one "disease" but several diseases, each having different properties, and that "resistance" is a relative, rather than absolute phenomenon, showing quantitative variation. The diseases generally considered important in barley are mildew, brown rust, leaf blotch and yellow rust and to a lesser extent, eyespot and barley yellow dwarf (BYDV). Other diseases of the crop, including net blotch, leaf stripe and halo spot, have been generally regarded as unimportant, and have been ignored by UK breeders unless a breeding nursery is infected when some selection could be practised. However, the recent increase in net blotch and the occurence of barley yellow mosaic virus on winter barley shows that these priorities require continual reassessment.

Quantitatively, resistance is best measured relative to that of other cultivars, rather than on any absolute scale, so resistance levels can be most easily considered with reference to a hypothetical spring barley cultivar "Newvar" soon to be submitted for National List testing (Table 1). For convenience, the rating systems for yield, agronomic characters and disease resistances used in the NIAB Recommended List for 1979 (Anon., 1979) are followed. The ratings of "Newvar" are approximate averages of the recommended cultivars and could, therefore, be readily achieved without intensive selection.

Even if it is successful, a new cultivar is unlikely to achieve more than 30% of the spring barley seed market (Table 2), and a more realistic expectation is 6-8% (giving a domestic annual income of around £250 000 from royalties and sales of basic seed). Since exceptional new cultivars are now often multiplied quickly for earlier marketing, and new breeding techniques have increased the rate at which improved germplasm is reused in new combinations, a cultivar is unlikely to remain on the Recommended List for more than 5 or 6 years, and this will also limit the expansion of its acreage. Commercially, the most logical breeding policy therefore rests on the production of a regular progression of cultivars showing modest improvements, rather than the occasional cultivar of outstanding merit which will be overtaken by others before it has fully realized its sales potential.

The first step in marketing a new cultivar is usually to procure its inclusion in the Recommended List because, although marketing is

Table 1. Projected characteristics of a hypothetical spring barley cultivar "Newvar" scheduled for release in 1980 compared with some established cultivars (Anon., 1979).

	Cultivar "Newvar"	Jupiter	Goldmarker	Georgie
Yield (% of Julia)	120	115	113	111
Average agronomic characters	7	6	8	7.5
Resistance to mildew	6	8	7	5
Resistance to yellow rust	4	3	6	3
Resistance to brown rust	4	4	3	5
Resistance to rhynchosporium	4	4	4	2
Year of release	1980	1975	1975	1974

possible without recommendation, it is rarely successful. Success also depends on how the cultivar is described on that list. The most important consideration is yield, which will clearly be affected by the presence of disease in trials. Since these trials are done in areas representative of the national crop, and "Newvar" has resistance levels similar to the averages of the other cultivars in current commercial use, it can be expected to suffer yield losses due to disease similar to those estimated for the national crop by the Ministry of Agriculture disease surveys, i.e. about 10%, attributable mainly to mildew (King, 1972, 1977; Cook *et al.*, this volume). Thus complete resistance to leaf diseases could raise its yield to over 130% of the controls, whereas

Table 2. Seed sales of the predominant spring barley cultivars in the UK in each year from 1967-79. (Percentage of total sales)

1967	Proctor	19	Impala	11
1968	Zephyr	33	Proctor	14
1969	Sultan	25	Zephyr	23
1970	Sultan	29	Zephyr	20
1971	Sultan	20	Zephyr	14
1972	Julia	27	Vada	14
1973	Julia	29	Golden Promise	11
1974	Julia	23	Golden Promise	14
1975	Julia	16	Golden Promise	14
1976	Golden Promise	17	Mazurka	14
1977	Golden Promise	14	Mazurka	13
1978	Golden Promise	14	Aramir	9
1979	Golden Promise	17	Georgie	14

lower levels of resistance would result in concommitant reductions in recorded yield. The losses due to disease in winter barley are less well known but are unlikely to average less than 10%. However, the relative importance of the various diseases could well be different from that in spring barley.

Inclusion on the Recommended List also requires certain minimum standards of resistance to be met, and is more likely if the higher desirable standards are achieved (Doodson, 1976; Priestley, this volume) (Table 3). Whilst these standards are not immutable, the current minimum standards are relatively easy to achieve and there is obvious merit in avoiding extreme genetic vulnerability. These standards are as much concerned with the maintenance of the national average level of resistance as with the performance of specific cultivars and are inevitably influenced by the use of fungicides in commercial crops.

Table 3. Minimum and desirable standards of disease resistance for winter (W) and spring (S) barley cultivars on the NIAB Recommended List (Doodson, 1976).

Disease	Crop	Minimum standard	Desirable standard
Mildew	W & S	3	6
Yellow rust	W	4	5
	S	3	4
Brown rust	W & S	3	None specified
Rhynchosporium	W & S	2	None specified

Clearly, the widespread and increasing use of fungicides must have important repercussions on resistance breeding and it has even been suggested by some breeders and growers (Pearson, this volume) that resistance need no longer be a consideration in cultivar selection. However, fungicides are rarely totally effective in practice, so some resistance is needed to provide additional protection. Furthermore, although high grain prices may currently justify heavy fungicide usage, especially in high-yielding winter barleys, it seems probable that consumer pressure will soon lead to lowered prices for grain, whilst the cost of chemicals and their application (typically estimated at current prices as *c.* 5% of the value of the crop per application) will continue to rise. Thus in the longer term, fungicide use must become less attractive than at present and resistance will become more important than it appears now.

An equally important influence on the market value of resistance is the need for it as perceived by individual growers under their partic-

ular conditions. Thus growers in south-west Britain, for instance,
seem particularly concerned about leaf blotch and brown rust and favour
cultivars with resistance to these diseases. Elsewhere, there seems
less awareness of the need for resistance and a survey showed that most
cereal growers consider disease resistance to be less important than
quality, yield, standing power and earliness. Undoubtedly, many of
these growers put their faith in fungicides for crop protection but it
is also probable that they have failed to appreciate the amount of
resistance already present in the cultivars to which they are accust-
omed. However, breeders must also appreciate that growers' perceptions
of the desirable attributes of a cultivar may change and must not therefore be allowed undue influence.

The conflict between short-term and long-term considerations arises
repeatedly in planning the development of a breeding programme. Even
using specialized techniques to accelerate breeding, the time between
crossing and significant seed sales is typically 10 years and contin-
gency arrangements are necessary in the early generations to accommodate
possible changes in requirements which may subsequently occur. Most
programmes carry material in the early generations showing a wide range
of resistance and susceptibility; some material is included solely for
agronomic performance, irrespective of susceptibility, whilst in other
material some agronomic qualities are sacrificed in order to meet
specific resistance requirements. The decision to discontinue particular material may occur at any stage up to and even after release.

RESISTANCE BREEDING STRATEGIES

Having reviewed some of the factors which influence resistance breeding
policy, it is now possible to consider the strategies most appropriate
for use in competitive breeding programmes. The classification of
resistance as being either horizontal or vertical is largely one of
convenience, since it is now generally accepted that there is no clear
division between these types in nature (Nelson, 1978). Strategies for
the use of horizontal resistance may use this type alone, or with
additional support from vertical resistance for enhanced crop protection
(Jones & Clifford, 1978). In fact, in the European barley gene pool it
is difficult to avoid the presence of some vertical resistance.
Vertical resistance can also be used alone, but under these circumstances it is often (though not inevitably) rapidly overcome by the
emergence of new virulence in the pathogen. To increase its durability,
alternative proposals for the use of vertical resistance include:
1. Deployment in space or time so that specified resistances are
    only used in certain geographical areas or only in winter or
    spring crops;
2. Diversification schemes, in which farmers are advised to grow
    several cultivars with different resistances in different
    fields (Priestley, this volume);
3. Growing mixtures of cultivars having different resistances
    (Jeger *et al.*, this volume; Wolfe *et al.*, this volume);
4. Breeding multiline cultivars, which incorporate a range of

different resistances into a common genetic background (Browni
& Frey, this volume);

5. The development of multigene cultivars, in which two or more
    resistance genes are combined into a single cultivar, and whic
    can only be overcome by the development of complex virulence i
    the pathogen.

Since the diseases of barley differ considerably in many respects, the
strategies most appropriate for each disease are considered separately
below.

Barley yellow dwarf virus

The importance of BYDV has not been objectively determined and no
resistance standards have been imposed but there does appear to be a
market for one or two resistant winter barley cultivars for early sowin
in south and west Britain. However, there is probably no necessity for
all cultivars to show resistance. In spring barley the disease is less
apparent and there does not presently appear to be any real demand for
resistance, so most breeders would not concern themselves with this
problem. Since the main source of inoculum is grass, little selection
pressure is imposed by resistant barley cultivars and vertical resistance can be expected to provide effective and durable control. The
gene *Yd2*, derived from C.I. 3906 and other Ethiopian cultivars and
present in cv. Coracle, provides resistance to all isolates of the viru
against which it has been tested.

Eyespot

A minor disease, eyespot is potentially of importance only in the winte
crop, and most cultivars appear to be moderately resistant (P.R. Scott,
personal communication). As greater levels of resistance are not
readily available, there seems little point in actively breeding for
resistance. However, some germplasm appears rather more susceptible,
and confirmatory testing of advanced material may be advisable to ensur
that resistance has not been lost.

Yellow rust

National disease surveys (King, 1972, 1977) suggest that, of the
diseases for which there are minimum resistance standards, yellow rust
is the least important and does not merit intensive resistance breeding
However, the disease undoubtedly has considerable epidemic potential,
and severe damage has sometimes been observed on very susceptible
cultivars. Thus it appears advisable to meet the minimum standard of
resistance, but there is little economic advantage in exceeding it.
In spring barley, a resistance rating of 4 can be readily achieved
through an apparently horizontal form of resistance present for example
in cvs Proctor, Lofa Abed and Zephyr. Better resistance, to a rating
of 6, can be obtained from cv. Arabische and its derivatives including
cvs Emir, Aramir, Athos and Hassan. Although this resistance appears
to be simply inherited and therefore probably vertical, it has so far
proved durable in practice. The introduction of additional sources of

yellow rust resistance, whilst an important national precaution, is not
an essential component of commercial programmes in the UK at present,
especially since improved sources may be introduced from time to time
in cultivars bred in countries where losses are greater. However,
although there appears to be little difficulty in maintaining adequate
resistance, a number of high-yielding cultivars released recently
(derived primarily from cvs Vada or Tellus) have shown alarming susceptibility and progeny from crosses involving them must be selected
with care.

Because the optimum temperature for growth of *Puccinia striiformis*
is relatively low, autumn-sown crops are potentially more vulnerable to
yellow rust than are spring-sown and higher standards of resistance are
required in winter cultivars to contain the disease. It is not known
whether the resistance in the currently grown winter barley cultivars
is horizontal or vertical but in general it appears fairly durable, so
there should be no difficulty in meeting the appropriate standard.
However, the introduction of some new sources of resistance into winter
barley can also be justified commercially.

Leaf blotch

Although the average national importance of leaf blotch in spring
barley is extremely low, it is a significant problem in north, south
and west Britain where occasional local epidemics in wet years can cause
severe damage. Within these areas, there is considerable awareness of
the problem and thus there is a recognized market, of up to 10% of the
national acreage, for two or three cultivars showing good levels of
resistance (ratings of 6 or over). This is an attractive commercial
proposition, since although horizontal resistance of this order is
available from cvs Proctor and Tyra, the evidence suggests that better
resistance and adequate durability can be conferred by the simple use
of vertical resistances. More generally, all cultivars should meet the
minimum standard of 2, which is readily achieved by horizontal resistance in all cultivars except those with the prostrate juvenile growth
habit. In the latter, only vertical resistances appear to provide
sufficient protection to counter the more rapid spread of disease
resulting from the prostrate habit. At present, only the genes *Rh*
(in cv. Armelle) and *Rh4* (in cv. Magnum) are used in commercial
cultivars and, whilst these can be expected to show considerable
durability, it may be advisable to introduce one or two additional
vertical resistances. However, although such a programme is in the
national interest, it is speculative on the decreased effectiveness of
Rh and *Rh4* and is therefore unlikely to justify use of resources in a
commercial programme.

Leaf blotch is much more damaging in winter barley and is potentially
the most important disease of this crop. Although most serious in the
north, south and west, it can be damaging throughout the country and
there is thus a considerable market for cultivars with a resistance
rating of 6 or more. Despite the widespread use of fungicides, climatic
conditions during autumn and winter are so favourable for disease devel-

opment that there seems a clear economic justification for breeding all
winter cultivars to a standard of 4, and several to a standard of 9.
The *Rh* gene, which appears to be present in most commercial cultivars,
is ineffective against race UK2 of *Rhynchosporium secalis* and, although
additional resistances appear to be present in most of these cultivars,
the introduction of further resistance genes should have a high
priority. The horizontal resistance found in adapted cultivars does
not appear adequate to protect the crop under the climatic conditions
encountered when it is sown in autumn.

Brown rust

Brown rust is most important on the spring crop especially in south and
west Britain, and growers in these areas place a high priority on resistance. Nationally, however, brown rust is of little importance and,
provided minimum standards are maintained, the market needs only two or
three cultivars with resistance ratings of 5 or more. Fortunately, such
resistance is readily available in cv. Vada and its derivatives (e.g.
cvs Georgie, Sundance and Lofa Abed), whilst many derivatives of cv.
Emir (e.g. cvs Aramir, Hassan and Tintern) easily meet the minimum
standard. It is not clear whether these resistances are horizontal or
vertical, but they both appear to be adequately durable and easily
transferred. The resistance of cv. Vada is frequently associated with
susceptibility to yellow rust, although also with resistance to mildew,
whilst the resistance of cv. Emir is often associated with resistances
to both yellow rust and mildew. Nationally, reliance on just two
sources of resistance is unwise, and the identification and introduction
of new resistances is desirable but, for the commercial programme, these
two resistances are quite adequate. Some recently released cultivars,
such as Triumph and Simon, are protected by major gene resistances which
are race specific and may thus cause problems in the future. Nevertheless, major gene resistance may be the only form with sufficient
expressivity to adequately protect cultivars of the *erectoides* type, a
character which appears to be pleiotropically associated with extreme
brown rust susceptibility.

   In winter barley, brown rust can be adequately suppressed by resistance of the minimum standard and although pathologists are concerned
that the pathogen may overwinter on susceptible crops, this cannot
realistically be expected to affect the policy of a competitive
programme.

Mildew

For all of the above diseases, the minimum and desirable standards of
resistance can be readily obtained with acceptable durability from
within the existing high-yield gene pool. Indeed, the hypothetical
cultivar "Newvar" could easily satisfy the disease resistance requirements for most of the barley growing areas, and it has not yet been
necessary to advocate any gene management scheme to assist in controlling these diseases. This is not so for mildew, which is by far the
most important disease of spring barley (Cook *et al.*, this volume), and

even cultivars with resistance ratings of 5 or 6 may be suffering average yield losses of around 6%, although over 50% of the acreage is treated with fungicides. It therefore seems essential that every new cultivar should have a minimum resistance rating of 6, and an enormous market (up to 50%) exists for cultivars with higher ratings. Moreover, higher resistance ratings are clearly advantageous in securing higher yield performance in Recommended List trials. Viewed pragmatically, there is no easier way to achieve a 6% yield increase, and hence probably top yield position in trials, than to incorporate a gene for vertical resistance to mildew into "Newvar". This practice has, of course, been recognized and followed by breeders for many years with considerable success; the problem has been the lack of durability of this resistance. Nevertheless, the production of successive cultivars with vertical resistance has undoubtedly prevented the considerable yield losses which would have occurred, given the pre-existing levels of susceptibility in the earlier cultivars.

However, most pathologists consider that greater durability of resistance is required, and it is probably to encourage the development of cultivars with durable resistance that the minimum resistance standard for mildew is maintained at the economically unsound level of 3 (on the assumption that durability could require horizontal resistance, which is likely to be expressed to a lower degree than vertical resistance). Although there are indications that a partial, non-hypersensitive type of resistance may be present in cvs Asse and Clermont and in several European land cultivars, the evidence that it is durable, its expressivity, and the agronomic characters of the source cultivars are all inadequate to justify the extensive use of this type of resistance in a commercial breeding programme. A realistic appraisal suggests that, whilst the presence of horizontal resistance in an agronomic background of adequate yield would be welcomed by the scientific community as meriting inclusion on the Recommended List, such a cultivar is not likely to be particularly acceptable to farmers. The likely durability of resistance is not a major consideration in their choice of cultivar.

Durability of resistance may alternatively be achieved by gene management. Unfortunately, schemes based on regional deployment of particular resistances, or on the use of different resistances in winter and spring crops, founder on two obstacles. Firstly, it is impossible to identify the genes present in all sources of resistance in use and thus some duplication of genes from different sources could not be avoided. Secondly, enforcement or co-operation is improbable since breeders are in competition and many cultivars used in Britain are bred elsewhere.

The use of diversification, either between fields (Priestley, this volume) or within fields as cultivar mixtures (Wolfe et al., this volume) offers more promise. However, it is difficult for breeders to plan sufficiently far ahead to design and develop their own schemes so these are better devised by the recommending authority using cultivars available at any given time. Thus a breeder can best safe-

guard his position in the market by producing as wide a range of
cultivars as practicable, thereby ensuring that at least some of his
cultivars are likely to fit into any diversification plan. Diversification through the breeding of multiline cultivars is essentially a
conservative approach, and is uncompetitive compared with less rest-
rictive breeding procedures.

Multigene, or at least plurigene, cultivars have not been extensively tested, largely because much of the genetic diversity available
is concentrated at a few loci, and because experience with other host/
pathogen systems (potato blight, wheat stem rust) has suggested that it
is ineffective. However, although certain gene combinations, for
example *Mla6/Mlg*, have lacked durability, others appear slightly more
promising, and the combination of various *Mla* resistances with the
resistance from cv. Vada is being used in many breeding programmes.

By contrast, it could also be suggested that the need for durability
of mildew resistance has been exaggerated. As even successful cultivars
occupy only a small proportion of the acreage, the rapid progress in
cultivar improvement in recent years has resulted in cultivars
frequently becoming obsolete before their resistance has been overcome
to any appreciable extent; there is no "boom" and therefore no "bust".
Furthermore, diversification is automatically assured by the failure of
any one cultivar to dominate the market, and fungicides are available to
protect cultivars with vertical resistance after this resistance has
become ineffective. Under these circumstances, the use of durable resistance confers little advantage to the individual breeder; it facilitates the continuing utilization of resistance in subsequent cycles of
crossing, but this benefit also accrues to the competing breeders. Thus
at present there is no effective commercial inducement to develop pure-
line cultivars with durable (i.e. horizontal) resistance. Of course,
this situation would change if a yield plateau were to be reached, if
fewer new cultivars were released, or if fungicide use declined.
Clearly, the identification and development of durable resistance is
necessary in the longer term, and is an important task for state
research programmes.

Since the hardest part of marketing a cultivar is getting it
established, a realistic strategy for commercial spring barley breeding
is to introduce an effective major gene for mildew resistance into a
high-yielding genotype which is protected to minimum standards for brown
rust, yellow rust and leaf blotch. This will ensure a high yield performance and acceptance on the Recommended List. By the time the mildew
resistance has begun to lose effectiveness, farmers' acceptance of the
cultivar will be such that many will continue to grow it, using fungicides or combining chemical control with gene management systems on the
farm. Meanwhile, additional resistance genes would be introduced into
new genotypes. Commercially profitable in its own right, this practice
of random introduction of new resistance genes is also fully compatible
with diversification strategies.

In winter barley, mildew is relatively less important, but still justifies considerable effort to incorporate resistance, following the same lines as proposed for the spring crop. Although it may be possible to suggest a gene management scheme to limit disease spread between winter and spring crops, this cannot be effectively done by breeders, but should follow recommendations by plant pathologists, based on the range of cultivars available.

CONCLUSIONS

Clearly, no single strategy can at present claim maximum disease control and minimum economic loss. On a national basis, therefore, safety lies in using a diversity of chemical, cultural and genetic control measures, including a diversity of resistance types, sources and strategies. Diversity of cultivars is best assured by maintaining a diversity of breeders, many of whom may hold views differing markedly from those expressed above. If we have stressed the necessity of short-term expediency over long-term provision, it is with the recognition that current progress in practical breeding is based on a foundation of research and development laid over a period of 20 years or more. It is clearly in the national interest, for future breeding progress, that such research should not be neglected, and one of the basic functions of state breeding organisations is to assist in this long-term development by identifying and exploiting new sources and mechanisms of resistance and by developing new breeding techniques which will make the best use of this new knowledge.

REFERENCES

Anon. (1979) *Recommended Varieties of Cereals*. National Institute of Agricultural Botany, Cambridge.
Doodson J.K. (1976) Disease standards and diversification of varieties as a means of decreasing cereal diseases. *Proceedings of the 14th National Institute of Agricultural Botany Crop Conference*, pp. 32-9.
Jones D.G. & Clifford B.C. (1978) *Cereal diseases: their pathology and control*. BASF, Ipswich. 279 pp.
King J.E. (1972) Surveys of foliar diseases of spring barley in England and Wales 1967-70. *Plant Pathology* 21, 23-35.
King J.E. (1977) Surveys of foliar diseases of spring barley in England and Wales 1972-75. *Plant Pathology* 26, 21-9.
Nelson R.R. (1978) Genetics of horizontal resistance to plant diseases. *Annual Review of Phytopathology* 16, 359-78.

The strategy of the International Maize and Wheat Improvement Center (CIMMYT) for breeding disease resistant wheat: an international approach

H. J. DUBIN* & S. RAJARAM†

*CIMMYT Andean Region Project, Quito, Ecuador
†CIMMYT Bread Wheat Project, El Batan, Mexico

INTRODUCTION

The CIMMYT wheat breeding programme aims to produce high yielding, broadly-adapted, disease-resistant germplasm for the less developed countries (LDCs). In 1967/68 high yielding, semi-dwarf wheat cultivars were grown on about 5M ha in LDCs; by 1976/77 this had increased to approximately 29M ha (Dalrymple, 1978). Most of these cultivars are either CIMMYT-derived or lines interbred with CIMMYT germplasm in national programmes. As the area sown with CIMMYT-related germplasm has increased, the need for broader-based and more stable disease-resistant cultivars has become more important.

This chapter describes the strategies used by CIMMYT to increase and stabilize disease resistance in bread wheat. Similar techniques are used in the triticale, durum wheat, and barley breeding programmes.

PRINCIPAL STRATEGIES

Introduction of diversity; stirring the genetic soup

Because wheat cultivars derived from CIMMYT are grown on such large areas and under very different conditions the germplasm used must be as diverse as possible. New lines must be developed quickly if they are to be of value to LDCs. Decisions are based on experience and a rapid assessment of the available data rather than on a detailed analysis of each cross. We must avoid the danger of "paralysis by analysis".

CIMMYT grows two generations and makes about 8000 - 10 000 crosses each year (4000 - 5000 crosses/generation). One generation is grown at Cuidad Obregon, Sonora (27°20'N, 40 m above sea level) and the other at Toluca (19°16'N, 2649 m above sea level); both sites are in Mexico. At Cd. Obregon wheat leaf rust (*Puccinia recondita*) and stem rust (*P. graminis*) are prevalent, whereas at Toluca the major disease is stripe rust (*P. striiformis*), although *Septoria nodorum*, *S. tritici*, *Fusarium nivale* and *F. roseum* occur sporadically.

Jenkyn J. F. & Plumb R. T. (1981) *Strategies for the Control of Cereal Disease*

In both generations all segregating and parental nurseries are inoculated with leaf and stem rust urediospores. Both locally prevalent rust isolates and those with uncommon virulence combinations are used. Border rows are sown with mixtures of universally and differentially susceptible lines to maintain as much diversity as possible in the pathogen.

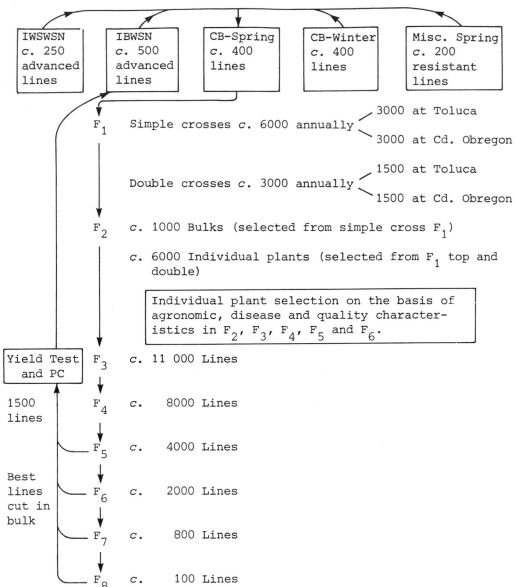

Figure 1. Management of parental material and progeny populations in the CIMMYT bread wheat breeding programme (IWSWSN = International Winter x Spring Wheat Screening Nursery, IBWSN = International Bread Wheat Screening Nursery, CB = Crossing Block, PC = Small Plot Multiplication of Pure Seed).

The germplasm used in the crosses is obtained from all areas of the world and chosen because it has some useful trait. Fig. 1 illustrates the movement and handling of materials in the CIMMYT bread wheat breeding programme. The numbers of lines differ from year to year but 40-50% of the plants selected from each generation are subsequently discarded due to unacceptable seed characteristics.

Since 1969, CIMMYT has been crossing spring with winter wheat on a large scale, in the hope that mixing these relatively isolated gene pools will increase potential yields. Spring wheat may also introduce leaf and stem rust resistance and increase the range of winter hardiness. On the other hand winter wheat may contribute resistance to drought, stripe rust and *Septoria* spp.. The International Winter x Spring Wheat Screening Nursery (IWSWSN) is derived from selections made at Oregon State University. Of the crosses made each year, 1000 - 1500 are spring x winter wheat. A complete description of this programme is given in CIMMYT Reports on Wheat Improvement (Anon., 1978, 1979, 1980).

As a result of many years experience of cereal breeding, CIMMYT places much emphasis on the pedigree breeding system and use of the double cross (Fig. 1). The double, or four way, cross permits rapid combination of germplasm from four parents in one year and increases the genetic diversity more quickly than other methods. Double crosses allow breeders to combine many genes for resistance to one disease at one time and also enable them to combine genes against several diseases rapidly. This is especially important because CIMMYT breeds for diverse climates and regions with many different disease problems.

Multilocation testing; CIMMYT's most important international strategy

Genetic studies have suggested that wheat genotypes resistant to rust diseases in many dissimilar localities, as indicated by low coefficients of infection (Anon., 1979), often contain multiple factors for resistance (Rajaram & Luig, 1972). However, it is not known whether some of this polygenic resistance is race-non-specific. Nevertheless, if a line contains several functioning resistance genes there is a greater chance that the resistance will be durable. By testing at a number of epidemiologically dissimilar sites, and exposing lines to the greatest possible range of virulence factors, the most durable and useful resistances should be identified.

Obviously such thorough screening cannot all be done in Mexico and F_2 bulk populations of the best simple crosses are sent for selection to co-operators in many other countries (Fig. 2). After several generations, the most promising advanced lines from each area are cycled back into the CIMMYT programme as parents.

Low average coefficients of infection indicate the presence of broadly-based resistance, and analyses of patterns of reaction to diseases at many sites with different virulence factor combinations allow lines with distinct resistance genes to be identified (Table 1). Although the individual genes cannot be identified by this method, it

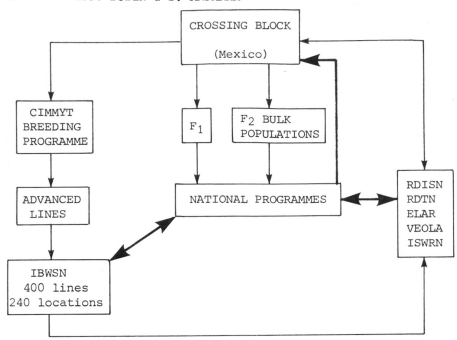

Figure 2. Flow of germplasm between national and CIMMYT breeding programmes (RDISN = Regional Disease and Insect Screening Nursery, RDTN = Regional Disease Trap Nursery, ELAR = Latin American Rust Trap Nursery, VEOLA = Latin American Disease and Observation Nursery, ISWRN = International Spring Wheat Rust Nursery, IBWSN = International Bread Wheat Screening Nursery).

does provide a simple, rapid means of identifying lines with different resistance genes for use in the breeding programme.

Recent leaf rust infection pattern analysis (Rajaram *et al.*, 1979) has shown that CIMMYT's Siete Cerros multiline components have many leaf rust resistance genes. In 11 crosses, representing 21 advanced line components, pattern analysis based on 30 localities has shown 16 different resistance groups to be present. Thus these components have at least 16 distinct leaf rust resistance genes. Seedling tests with leaf rust cultures confirmed the presence of at least seven known and three unknown genes.

Although the main aim of the programme is to combine genes from many areas, many crosses are made between lines from the same region to meet the needs of specific countries. The most resistant lines from different regions are then intercrossed to improve the resistance base and to provide more widely adapted germplasm.

CIMMYT considers leaf rust, stem rust and stripe rust to be most important. However, screening for resistance to other diseases is justified where they are especially common and economically important.

Table 1. Patterns of leaf rust infection on two advanced wheat lines at six sites.

	Leaf rust reaction*					
Advanced line	Natal S. Africa	Sonora Mexico	Madhya Pradesh India	Buenos Aires Argentina	Ancash Peru	Giza Egypt
Chiroca sib (10th IBWSN entry 106)	5R	10R	TS	0	TS	0
Moncho sib (10th IBWSN entry 63)	20S	20M	50S	10S	40MS	TS

* Modified Cobb Scale (see Chester, 1950).

At present, *S. tritici, S. nodorum, Erysiphe graminis, Drechslera* spp. and Barley Yellow Dwarf Virus come into this category.

The International Nursery System   Data from many sites are obtained through the International Nursery System. These international nurseries originate either in Mexico or from regional CIMMYT programmes. Fig. 2 shows how they link the various parts of the CIMMYT and national breeding programmes.

The principal nursery, which tests lines after several years selection and yield testing, is the International Bread Wheat Screening Nursery (IBWSN). This nursery, now in its 13th year of operation, tests 400-500 of the highest yielding and most disease resistant lines selected in the Mexico base breeding programme. Seed of these lines is distributed to about 240 centres throughout the world. Data are received from about 36% of the centres and used to calculate average coefficients of infection for specific diseases. The best lines are then cycled back into the CIMMYT base breeding programme in Mexico.

There are two disease screening nurseries which evaluate lines in greater detail. The Regional Disease and Insect Screening Nursery (RDISN) is administered from Cairo, Egypt, in co-operation with the International Center for Agricultural Research in the Dry Areas (ICARDA) while the Latin American Disease and Insect Screening Nursery (VEOLA) is administered from Quito, Ecuador. The lines sown in the RDISN and VEOLA are obtained from national programmes in their respective regions and are the most disease resistant and agronomically desirable types from the eastern and western hemispheres. The lines with the most broadly-based resistances are available to all co-operating nations for use in their national programmes and are also used in the CIMMYT base programme (Saari, 1978).

Two nurseries, which are important in ensuring the stability of wheat production in their regions, are the Regional Disease Trap Nursery (RDTN) and the Latin American Rust Nursery (ELAR). These nurseries aim to detect changes in pathogen virulence as early as possible, and so provide early-warning to countries of the need to remove newly susceptible cultivars from production and substitute more resistant ones. The nurseries are sown with commercial cultivars, advanced lines being multiplied, iso-genic differentials for the three wheat rusts and some cultivars to test for other diseases (Prescott et al., 1976).

These nurseries are usually sown in areas where there are many different pathogen virulence genes and in representative wheat growing areas of the region.

The International Spring Wheat Rust Nursery (ISWRN), which started over 28 years ago, is administered by the United States Department of Agriculture (USDA) and is another excellent source of disease resistant germplasm for CIMMYT and the developing nations.

SUPPORTIVE STRATEGIES

Multiline composites

Another way to achieve greater stability of wheat yields is through the development and use of multiline composites. This means of manipulating rust resistance and the methods used by CIMMYT are discussed in greater detail elsewhere (Rajaram & Dubin, 1977; Rajaram et al., 1979) but Table 2 shows the main differences between the current CIMMYT approach to producing multiline composites and the classical approach. The CIMMYT approach is considered more practical and useful, especially as it suits the needs of the LDCs.

CIMMYT's responsibility is to *produce* the components for a multiline and over the last 8 years more than 500 cultivars or lines have been used in top or double crosses with cv. Siete Cerros to produce components of the 8156 multiline. It is the responsibility of the pathologists and breeders in each country to *evaluate and select* the most useful lines for their own areas.

In 1978 a multiline composite (Bithoor) was released in India. It consists of nine 8156 components bred at CIMMYT but selected in India. This is the first multiline derived from 8156 to be released.

Table 2. Differences between multilines produced using the CIMMYT and classical approaches.

CIMMYT Approach	Classical Approach
Products of three or four way crosses	Products of backcrossing
One to many effective genes in each component	Single effective gene in each component
Rapid production of components	Slow and relatively laborious production of components
All components resistant to all rust strains	Specific components susceptible to certain rust strains
Composite is a combination of phenotypically similar lines	Components are near-isogenic lines
Heterogeneous for non-target diseases	Homogeneous for non-target diseases
Individual components can serve as cultivars	Individual components should not serve as cultivars

Identification and utilization of slow rusting or dilatory resistance

Dilatory resistance (Browning *et al.*, 1977) which in relation to rusts is commonly known as slow rusting (Caldwell, 1968), is distinct from the classical, well understood and long utilized, hypersensitive type of resistance. Slow rusting has been little used in wheat breeding programmes but at CIMMYT we hope to identify and use this form of resistance together with pyramided, hypersensitive resistance to prolong the usefulness of cultivars. However, it should not be assumed that dilatory resistance is invariably race-non-specific or horizontal and much more research is needed to evaluate its nature properly (Ellingboe, 1976).

At CIMMYT, a simple procedure is used to identify lines apparently having this type of resistance to leaf rust. Any of the 1500-2000 advanced lines which yields well but is susceptible to leaf rust, is selected and urediospores of the compatible rust collected and multiplied. These lines are then tested using methods similar to those of

Wilcoxson *et al.* (1975) and those lines which consistently exhibit slow rusting are used in the crossing programme. In F_2 populations, plants with rust reactions less than 10S may also be retained.

Torim 73 is a slow rusting cultivar which is grown widely in the Yaqui Valley, Sonora, Mexico. Although it is susceptible to several races of leaf rust which were epidemic on cv. Jupateco 73 in this region in 1976-77, fields of cv. Torim 73 rarely suffered any significant yield loss even when inoculum levels were very high. Torim 73 has been widely grown for more than 5 years and the resistance continues to be effective.

Incorporation of alien resistance

A small effort is made to incorporate alien resistance genes already in wheat backgrounds into the CIMMYT resistance-gene pool. Lines containing these genes are crossed with susceptible, high yielding, well adapted cultivars in backcross programmes, using the latter as the recurrent parents. When the progeny are stable the best ones are used in the normal double cross programme.

CONCLUSIONS

The best way in which CIMMYT can provide the LDCs with high yielding wheat lines adapted to a wide range of conditions and with resistance to many pathogen races, is through a large double cross breeding programme, associated with multilocation testing. This is CIMMYT's principal strategy but multilines, dilatory resistance and alien genes are also used to augment resistance. It is essential that there is a free flow of germplasm between countries if these aims are to be achieved.

REFERENCES

Anon. (1978) International Maize and Wheat Improvement Center. *CIMMYT Report on Wheat Improvement 1976*. El Batan, Mexico. 234 pp.

Anon. (1979) International Maize and Wheat Improvement Center. *CIMMYT Report on Wheat Improvement 1977*. El Batan, Mexico. 245 pp.

Anon. (1980) International Maize and Wheat Improvement Center. *CIMMYT Report on Wheat Improvement 1978*. El Batan, Mexico. (In press.)

Anon. (1979) *Results of the International Spring Wheat Rust Nursery 1977*. United States Department of Agriculture, Beltsville, Maryland, U.S.A. 244 pp.

Browning J.A., Simons M.D. & Torres E. (1977) Managing host genes: epidemiologic and genetic concepts. *Plant Disease: An Advanced Treatise, Vol. 1* (Ed. by J.G. Horsfall & E.B. Cowling), pp. 191-212. Academic Press, New York.

Caldwell R.M. (1968) Breeding for general and/or specific resistance. *Proceedings of the Third International Wheat Genetics Symposium, Canberra, Australian Academy of Sciences*. pp. 263-72.

Chester K.S. (1950) Plant disease losses: their appraisal and interpretation. *Plant Disease Reporter Supplement* <u>193</u>, 189-362.

Dalrymple D. (1978) Development and spread of high yielding varieties of wheat and rice in the less developed nations. *United States Department of Agriculture. Foreign Agricultural Economic Report* <u>95</u>, 134 pp.

Ellingboe A.H. (1976) Genetics of host-parasite interactions. *Physiological Plant Pathology* (Ed. by R. Heitefuss & P.H. Williams) pp. 761-78. Springer-Verlag, Berlin.

Prescott J.M., Saari E.E., Stubbs R.W., Boskovic M. & Kamel A.H. (1976) Summary report of the Regional Disease Trap Nursery (RDTN) 1974-5. *Mimeographed Report. Wheat Research and Training Project, Ankara, Turkey.* 66 pp.

Rajaram S. & Dubin H.J. (1977) Avoiding genetic vulnerability in semi-dwarf wheats. *The Genetic Basis of Epidemics in Agriculture* (Ed by P.R. Day). *Annals of the New York Academy of Sciences* <u>287</u>, 243-54.

Rajaram S. & Luig N.H. (1972) The genetic basis for low coefficient of infection to stem rust in common wheat. *Euphytica* <u>21</u>, 363-76.

Rajaram S., Skovmand B., Dubin H.J., Torres E., Anderson R.G., Roelfs A.P., Samborski D.J. & Watson I.A. (1979) Diversity of rust resistance of the CIMMYT Multiline Composite, its yield potential, and utilization. *The Indian Journal of Genetics and Plant Breeding* <u>39</u>, 60-71.

Saari E.E. (1978) Wheat disease surveillance and its role in genetic vulnerability. *Workshop on Crop Surveillance and Pest Management. East-West Center, Hawaii, U.S.A.*

Wilcoxson R.D., Skovmand B. & Atif A.H. (1975) Evaluation of wheat cultivars for ability to retard development of stem rust. *Annals of Applied Biology* <u>80</u>, 275-81.

The multiline concept in theory and practice

J. ARTIE BROWNING & K. J. FREY
Iowa State University, Ames, Iowa 50011, U.S.A.

INTRODUCTION

A strategy for cereal disease control that pathologists and breeders should consider more is the use of diversity. In biology, as in finance, diversity is a hedge against future risk and this strategy is, therefore, especially well suited for use against diseases caused by highly variable pathogens.

A multiline cultivar is a mixture of isolines that differ by single, major genes for reaction to a pathogen. Thus, multilines are a convenient way for using diversity in commercial crops, for which agronomic uniformity is desired.

Exactly how diversity in the host population effects disease control is not clear but it is plainly evident that the mixing of cultivars and species has such complex effects (Simmonds, 1962) that it is necessary to study isogenic lines to learn the role of diversity *per se* in buffering against pathogens. This chapter will discuss multiline cultivars that have a uniform genetic base and draw on over two decades of research and experience obtained with the multiline programme in Iowa.

THE MULTILINE CONCEPT

Cultivar development in autogamous crop species has proceeded by stages: introduction, development of land cultivars, pure line selection from land cultivars and pure line selection after hybridization. At each stage the genetic base of the crops has been progressively narrowed. The widespread use of hybridization for cultivar development has increased the frequency and intensity of epidemics, probably because it has led to the use of single resistance genes in cultivars grown over large areas or even whole continents. Thus, epidemics caused by a pathogen strain able to parasitize cultivars carrying particular resistance genes can develop over large areas. Almost invariably, some pathogen has adapted to each new, extensively planted host genotype and caused unprecedented epidemics.

Epidemics are not new; it is their intensity, extent, and frequency that are new. It is of no advantage to the pathogen to kill its host, but as it does not have evolutionary time to adapt to the new situation, it responds to its existing genetic information by reproducing at the rate which was necessary to *maintain* itself in the dynamically balanced indigenous ecosystem where it coevolved with the host, neither eliminating the host nor itself (Browning *et al.*, 1979). However, on host genotypes with uniform susceptibility this natural rate of reproduction has the potential to destroy popular host cultivars and hence the pathogen, if the host were not maintained by man. This has led to the "boom and bust" cycles so common among cereal cultivars and caused widespread disenchantment with that valuable natural resource, vertical resistance (VR).

During the 1940s and 1950s, when wheat and oat cultivars were protected from diseases only by single VR genes, the frequency and intensity of epidemics increased. This led to alternative suggestions for management of disease resistance genes, including their use in diverse host populations. Rosen (1949), for example, suggested using the bulk population from an oat cross, which would be heterogeneous for resistance to crown rust (*Puccinia coronata* var. *avenae*) and Victoria blight (*Helminthosporium victoriae*). Jensen (1952) proposed "intra-varietal diversification" by blending similar, but not isogenic, lines. Borlaug (1953) and Borlaug & Gibler (1953) advocated blending backcross-derived wheat lines into a multiline cultivar.

In Mexico, Borlaug (1959) began multiline breeding early in his Rockefeller Foundation (later CIMMYT) programme. However, his breeding programme also produced pure line cultivars with photoperiod insensitivity that outclassed previous agronomic types, such as the "Yaqui 50" multiline, so no multiline was released from the Mexican programme.

However, two multiline cultivars of wheat, "Miramar 63" and "Miramar 65", were released from the Rockefeller Foundation programme in Colombia (Anon., 1963-1965). This pioneering programme, like Borlaug's, was large and involved 2 or 3 backcrosses onto the recurrent parent (cv. Frocor) and use of over 600 donor wheats to develop 1,200 isolines, 10 of which were mixed equally to produce "Miramar 63" (Anon., 1963). Such large programmes have been criticized (Caldwell, 1966). However, this one was successful. These two pre-semidwarf multilines were well received by growers, controlled stripe and stem rust as they were designed to do, and are still in use in the high, cool Andes of Colombia.

CIMMYT's interest in multilines was only latent. After the semi-dwarf type "8156" proved successful in commercial production on millions of hectares around the world under the names "Siete Cerros" (Dubin & Rajaram, this volume) "Mexipak", "Kalyansona", etc., it was used as the agronomic background on which multiline cultivars produced in national breeding programmes were based (Rajaram & Dubin, 1977). Two 8156-type multilines with resistance to stripe rust and one with a different back-

ground were released recently from the national programme in India (Gill
et al., 1979; M.V. Rao, personal communication). Also, a stripe rust-
resistant wheat multiline "Tumult", was released recently in the
Netherlands (Groenewegen & Zadoks, 1979) and a barley multiline with
resistance to powdery mildew is being developed in Denmark (J.H.
Jørgensen & J.E. Hermansen, personal communication).

Thus, the pressing need of 30 to 40 years ago to control highly
epidemic cereal diseases has not been solved with pure-line management
of disease resistance genes, and interest has returned to multilines.
Evidence of this is the recent increase in the number of multiline
cultivars released and in the literature (especially on modelling), and
the organization of an international symposium "Use of Multilines for
Reducing Disease Epidemics" held in New Delhi, India, in 1978 (Anon.,
1979). In addition, the 1970 southern corn leaf blight epidemic (caused
by Helminthosporium maydis race T), which was attributed to the genetic
uniformity of the corn (maize) cultivars grown in the USA, has generated
new interest in heterogeneity (Day, 1977).

The history and theory of multiline cultivars was reviewed by
Browning & Frey (1969). Multilines affect the interaction between host
and pathogen populations both epidemiologically and genetically. As
individual host lines possess different major genes for resistance, they
decrease the effectiveness of incoming inoculum (I_i) and this results in
less initial disease (X_O), as is characteristic of VR (Vanderplank,
1968). Further, if a susceptible plant becomes infected, some of its
neighbours will probably be resistant, which decreases the probability
that inoculum from it will cause secondary infection. The apparent
infection rate (r) is therefore decreased and the progress of the
epidemic slowed in a way characteristic of horizontal resistance (HR)
(Vanderplank, 1968). Thus, genes for VR, managed in mixtures of isolines, can give the epidemiological effects that Vanderplank (1968)
attributed to both VR and HR.

These decreases in r and X_O seemed to explain the quantitative
epidemiological benefits of growing multiline cultivars. However,
Fried et al. (1979) recently presented data to show that multilines do
not reduce r, and Johnson & Allen (1975) suggested induced resistance as
a possible mode of multiline action.

THE IOWA MULTILINE PROGRAMME

The Iowa multiline programme began in 1957 (Frey et al., 1971) as part
of a comprehensive gene-management system against crown rust of oats
(Frey et al., 1977). We chose as recurrent parents two agronomically
proven oat lines, one early and one midseason in maturity. Isolines
were developed in each background by using a backcrossing programme.
Each recurrent parent was crossed with carefully chosen donor lines that
carried unique and useful genes for crown rust resistance. After an
original cross, five backcrosses (BC) were deemed necessary because many
donors were of unadapted lines or weed species, and because the public

had not then been educated to accept slight phenotypic variation. BC_5F_2 plants from each line of descent were rated for resistance and BC_5F_3 progeny rows from a line of descent that were homogeneous for resistance and that conformed to the agronomic type of the recurrent parent were bulked to form one isoline. Resistance genes were derived from *c*. 15 lines of *Avena sativa*, *c*. 11 collections of *A. sterilis* and one derived tetraploid.

Three sources of information guide our choice of isolines to include in Breeder Seed of a multiline cultivar.

1. Tests of agronomic performance. As crown rust is not serious every year, a multiline cultivar must yield as well in non-epidemic years as a good pure line cultivar. Therefore, all isolines are tested extensively in a range of environments and only isolines equal or superior to the recurrent parent in yield, weight per unit volume, lodging resistance, and resistance to other diseases are used in multiline cultivars.
2. Single-race rust nurseries. All isolines used in current multiline cultivars, candidate isolines, and recurrent and donor parents are tested annually in the adult stages with 10-12 crown rust races. The test races are those that are prevalent in North America or have the potential to become prevalent.
3. USDA annual continental crown rust virulence survey. This survey detects new virulence gene combinations and probable virulence trends.

Table 1 shows the isoline composition of a typical multiline cultivar Multiline E74. Seed of each isoline is maintained separately. For Multiline E74, 4.5 tonnes of Breeder Seed was composited to produce Foundation Seed. Registered and Certified Seed are produced and marketed by private industry. Seed cannot be certified if it is more than three generations removed from Breeder Seed. As new Breeder Seed must be composited each year it is easy to modify the isoline composition of a multiline cultivar. We have released eight early and five midseason multiline cultivars since 1968.

Currently we are developing isolines in a background of cv. Lang (Brown & Jedlinski, 1978), which matures early. To transfer useful genes from existing isolines we expect to need only three backcrosses as all parents are of good agronomic type, unlike those in our initial programme.

The contribution of multilines to control of crown rust in Iowa

This is difficult to assess. Iowa is a small part of the vast epidemiological unit that is the Puccinia Path of central North America. As oat in the Corn Belt were gradually replaced by corn and soybeans, the oat area in Iowa declined from *c*. 2 M ha in 1957 when we began our multiline programme to 0.7 M ha by 1968 when Multiline E68 and Multiline M68 were released. Very quickly, and for several years thereafter, an estimated 0.4 M ha were sown to multilines. Later, however, the proportion sown

Table 1. The oat isolines, their pedigrees, reactions to key races of *Puccinia coronata* var. *avenae*, and the percentage of each included in Breeder Seed of Multiline E74 (from Frey *et al.*, 1977).

CI no.	Pedigree	Reaction* to crown rust biotype			Percentage contribution in Multiline E74
		290	326	264B	
9169	CI $8044^4$2 x CI 7555^4 x Ceirch du Bach	MR	S	MS	18
9170	CI $8044^5$3 x Clinton x Garry 2 x CI 8079†	R	R	R	22
9172	CI 8044^6 x CI 8001†	MR	MS	MS	4
9176	CI 8044^6 x Acencao	R	MR	S	4
9177	CI $8044^4$3 x Clintland 2 x Chapman 178 x CI 7235	MR	MS	MS	4
9178	CI $8044^4$3 x Bonkee 2 x CI 7154 x CI 7171	R	MS	MR	18
9179	CI $8044^4$2 x CI 7555 x CI 6665	MS	S	MS	4
9180	CI $8044^4$2 x CI 7555 x CI 7654	MR	MR	MS	4
9181	CI $8044^5$2 x Clinton x CI 8081†	MR	R	R	22
Total percentage protection (R + MR + MS) from each race		100	78	96	–

*R, Resistant; MR, Moderately resistant; MS, Moderately susceptible; S, Susceptible.

†*Avena sterilis*

to multilines fell and by 1979 it was only 10% of the total of 0.6 M ha sown to oats. No damage by crown rust has been reported or observed on the multiline cultivars during their 13 yr of use.

The decrease in use of multilines can probably be attributed, in part, to their success in controlling crown rust and hence decreasing inoculum pressure in areas where they are grown. However, inoculum pressure was also diminished by the decrease in the total area sown to oats in Iowa and to the south and by the virtual eradication from Iowa of *Rhamnus cathartica*, the alternate host of *P. coronata*. Experiments in the early 1970s using paired sprayed and infected plots in 4 ha fields of a recommended, but susceptible, cultivar occasionally demon-

strated losses of 30%; in similar experiments with multilines, no loss
was detected. Thus, crown rust remains a threat when inoculum is
present and the environment is favourable. We are convinced that the
multiline cultivars were a main cause of the reduction in crown rust in
Iowa, and it seems ironical that their success has decreased the need
for them. Also, these multiline cultivars gradually became agronomically inferior to new pure-line cultivars with increased general crown-
rust resistance and tolerance which enabled them to perform adequately
against the decreased crown-rust pressure. It is possibly of more
importance that our existing multiline cultivars are susceptible to the
barley yellow dwarf virus (BYDV). Since the USA pandemic of BYDV in
1959, residual infection of perennial grasses has probably provided
inoculum for the infection of cereals, causing losses that went unrecognized. Pure-line cultivars which are tolerant to BYDV, such as Lang,
have yielded best in recent years. The new Lang multiline that we are
creating currently will combine crown rust and BYDV resistance, and it
should be the USA cultivar least vulnerable to these two principal
diseases of oats.

Epidemiological and resistance studies with Iowa multilines

Results from epidemiological studies have corroborated our observations
on the effectiveness of multiline cultivars in commercial production
against crown rust. In a typical experiment, Cournoyer (1970) trapped
crown rust spores quantitatively from the perimeter of isolated field
plots (16 x 16 m) to give an objective estimate of rust increase within
the plots. She compared disease progress by plotting cumulative spore
yields. When subjected to a severe epidemic (induced artificially using
four crown rust races that included race 264, to which three out of
eight isolines, representing *c.* 45% of Multiline M68, were resistant or
moderately resistant), the multiline produced only 25% as many spores as
the susceptible check isoline. Grain yield of the multiline was 10%
less than that of the resistant check, and that of the susceptible
isoline 50% less.

After multiline protection from crown rust had been demonstrated in
the short (4 wk) disease season in Iowa, multilines were grown on the
Texas coastal plain, where crown rust overwinters, to test their
effectiveness in a long (4 month) disease season. The experiments were
near Robstown and Beeville, Texas (*c.* 28° N.). Inoculum was from
natural, extraneous sources and very severe epidemics occurred in the
large (*c.* 0.5 ha) isolated field plots.

Relative urediospore yields from multilines and susceptible isoline
checks were the same in Texas as in Iowa, thus showing that multilines
will protect oats from crown rust in both short and long disease
seasons, even in very favorable environments.

The Texas experiments allowed us to estimate the proportion of
resistant plants needed to afford adequate protection in a diverse host
population. One metre indicator rows of component isolines and other
lines planted alongside the large plots at Robstown showed that,

although the multilines were *c.* 50-60% susceptible to some component of the natural pathogen population, they gave good protection. By contrast, mixtures containing 50% of a single resistant and 50% of a single susceptible pure line developed severe epidemics of crown rust. Thus, a severe rust epidemic can develop if 50% of the plants in a population are susceptible providing the susceptibility is uniform, not as in multilines.

At Beeville, where the environment was less favorable for crown rust, it still killed the susceptible pure line check prematurely except in plots protected with 11 weekly applications of fungicide. By contrast, the multiline and a mixture in the ratio of one resistant to two susceptible pure lines had no more rust than did susceptible, fungicide-sprayed plots. Thus, in this environment, when only one third of the plants were resistant, protection was equivalent to that given by 11 fungicidal sprays, even though the resistance and susceptibility were of uniform types!

The indicator rows at Robstown showed that the pathogen population was more diverse in the multiline than in the pure lines. Preliminary analysis of urediospores collected from pure and multiline plots in Texas confirm that spore populations from the multilines have more virulence genes in more combinations than those from pure lines. Significantly, however, the more diverse pathogen population has not resulted in an increase in rust. Wolfe & Barrett (1980; this volume) obtained similar results using mixtures of three barley cultivars to control powdery mildew.

CONCLUDING REMARKS

Multiline cultivars are but one method of achieving diversity. They are especially suitable where there are only one or two highly epidemic diseases, adequate genes for VR are available, and there is a requirement for agronomic uniformity.

Caldwell (1966) criticized the multiline method as being agronomically conservative and requiring considerable resources, but our multiline development programme has been small. Indeed, the proportion of our total effort devoted to breeding for crown rust resistance was decreased by the change to the multiline development programme. All backcrossing and routine rust testing are delegated to technical staff. Considerable project resources and scientific leadership are therefore freed for other activities, including creating inherently better disease management systems (Browning, 1974; Frey *et al.*, 1977) and "inherently better crop varieties" (Caldwell, 1966). We, like Groenewegen & Zadoks (1979), have obtained new genotypes as byproducts of our multiline programme. For instance, we found an association between factors for increased yield (up to 10%) and a crown rust resistance gene (Frey & Browning, 1971). It was partly because there was no need to fight diseases that breeding breakthroughs such as high protein lines (Frey, 1977) and yield increases of up to 25% (Frey, 1976) were achieved. One is free to use such programme "bonuses" either as parts of multilines or as pure lines.

However, everyone does not agree with our philosophy of controlling disease damage to crops by diversity. Several theoretical models, which investigate the problem of host-pathogen dynamics in multilines, have been published. In general, conclusions from these models do not agree with our experience with multilines or with natural ecosystems, but there are a number of possible explanations. For example, estimates of X_O and r were sometimes based on counts of uredia in small, inadequately isolated plots. This technique involves handling plants, which unwittingly supplements air dispersal of rust spores and facilitates interplot contamination. This method, therefore, underestimates the role of resistant tissue in the population and predicts more disease than we experience and that a greater proportion of resistant plants is required than we find necessary.

Most models also accept that stabilizing selection *sensu* Vanderplank (1968) is universal and enter a cost in reduced fitness for each virulence gene. A prior concept of stabilizing selection is that of Schmalhausen, cited by Lerner (1954) as basic to genetic homeostasis. In this view, stabilizing selection is "the rejection by natural selection of the extreme deviates . . . from the norm." Stabilizing selection *sensu* Vanderplank (1968) does not allow for exceptions (Nelson, 1972); natural selection to maximize the average, not the extremes (such as super races), does. Pathogens are frequently dis- cussed as if they have unique genetic systems. However, survival of traits (such as virulence and aggressiveness) in pathogen populations is governed by the same laws of genetic homeostasis that govern other organisms (Browning, 1978). In the history of pure-line cultivars, "bust" cycles have occurred because *any* race that can attack that cul- tivar is a super race. Given enough host diversity, however, the pathogen's self regulatory mechanisms operate so that super races need not be feared. Our results from Texas, those reported by Wolfe & Barrett (1980) for barley powdery mildew, and results from indigenous ecosystems (Browning, 1974) all support the conclusion that with diversity super races are unlikely to occur.

Finally, models fail to consider epidemiological units. For example, multilines in Iowa, a small area in the vast Puccinia Path, can hardly be expected to influence the race population in the entire epidemiological unit, much less generate a super race.

Regardless of predictions, and whether diversity reduces X_O, r, or affects epidemics in other, yet unknown ways, our multilines have given dramatic control of crown rust in Iowa and South Texas. Furthermore, our data are corroborated by data from the same pathogens and their wild hosts in indigenous ecosystems in Israel (Browning, 1974). This emphasizes the naturalness of using diversity and its promise for permanence.

There is potential for still more diversity, as Groenewegen &

Zadoks (1979) recognized in their plea for "*Poly-genotype* varieties: The multilines of the future." Meanwhile, a useful compromise between the needs for diversity and the demands for agronomic uniformity is provided by cultivar mixtures. Wolfe & Barrett (1980; this volume) using three-way barley cultivar mixtures, also obtained "dramatic" control of a highly epidemic pathogen.

ACKNOWLEDGEMENT

We gratefully acknowledge helpful suggestions on the manuscript by Christopher C. Mundt.

REFERENCES

Anon. (1963) Program in the Agricultural Sciences. *Report of the Rockefeller Foundation for 1962-63*, pp. 73-84.

Anon. (1965) Program in the Agricultural Sciences. *Report of the Rockefeller Foundation for 1964-65*, pp. 59, 70-2.

Anon. (1979) Use of multilines for reducing disease epidemics. *Indian Journal of Genetics and Plant Breeding* 39, 1-109.

Borlaug N.E. (1953) New approach to the breeding of wheat varieties resistant to *Puccinia graminis tritici*. *Phytopathology* 43, 467. (Abstr.)

Borlaug N.E. (1959) The use of multilineal or composite varieties to control airborne epidemic diseases of self-pollinated crop plants. *Proceedings of the First International Wheat Genetics Symposium, Winnipeg*, 1958, pp. 12-26.

Borlaug N.E. & Gibler J.W. (1953) The use of flexible composite wheat varieties to control the constantly changing stem rust pathogen. *Agronomy Abstracts*, p. 81.

Brown C.M. & Jedlinski H. (1978) Registration of Lang oats. *Crop Science* 18, 525.

Browning J.A. (1974) Relevance of knowledge about natural ecosystems to development of pest management programs for agro-ecosystems. *Proceedings of the American Phytopathological Society* 1, 191-9.

Browning J.A. (1978) Genetic homeostasis and the survival of traits in populations of fungal parasites. *Abstracts of Papers. 3rd International Congress of Plant Pathology, Munich*, 1978, p.92.

Browning J.A. & Frey K.J. (1969) Multiline cultivars as a means of disease control. *Annual Review of Phytopathology* 7, 355-82.

Browning J.A., Frey K.J., McDaniel M.E., Simons M.D. & Wahl I. (1979) The bio-logic of using multilines to buffer pathogen populations and prevent disease loss. *Indian Journal of Genetics & Plant Breeding* 39, 3-9.

Caldwell R.M. (1966) Advances and challenges in the control of plant diseases through breeding. *U.S. Department of Agriculture ARS* 33-110, 117-26.

Cournoyer B.M. (1970) Crown rust epiphytology with emphasis on the quantity and periodicity of spore dispersal from heterogeneous oat cultivar-rust race populations. Ph.D. Thesis, Iowa State University.

Dissertation Abstracts 31, 3104-B.

Day P.R. (1977) The Genetic Basis of Epidemics in Agriculture. *Annals of the New York Academy of Sciences* 287, 1-400.

Frey K.J. (1976) Plant breeding in the seventies: Useful genes from wild plant species. *Egyptian Journal of Genetics and Cytology* 5, 460-82.

Frey K.J. (1977) Protein of oats. *Zeitschrift für Pflanzenzüchtung* 78, 185-215.

Frey K.J. & Browning J.A. (1971) Association between genetic factors for crown rust resistance and yield in oats. *Crop Science* 11, 757-6(

Frey K.J., Browning J.A. & Grindeland R.L. (1971) Implementation of oat multiline cultivar breeding. IAEA-PL 412/17. *Mutation Breeding for Disease Resistance. International Atomic Energy Agency STI PUB/271,* Vienna, 159-69.

Frey K.J., Browning J.A. & Simons M.D. (1977) Management systems for host genes to control disease loss. *Annals of the New York Academy of Sciences* 287, 255-74.

Fried P.M., MacKenzie D.R. & Nelson R.R. (1979) Disease progress curves of *Erysiphe graminis* f. sp. *tritici* on Chancellor wheat and four multilines. *Phytopathologische Zeitschrift* 95, 151-66.

Gill K.S., Nanda G.S., Singh G & Aujla S.S. (1979) Multilines in wheat - A review. *Indian Journal of Genetics and Plant Breeding* 39, 30-7.

Groenewegen L.J.M. & Zadoks J.C. (1979) Exploiting within-field diversity as a defense against cereal diseases: A plea for "poly-genotype" varieties. *Indian Journal of Genetics and Plant Breeding* 39, 81-94.

Jensen N.F. (1952) Intra-varietal diversification in oat breeding. *Agronomy Journal* 44, 30-4.

Johnson R. & Allen D.J. (1975) Induced resistance to rust diseases and its possible role in the resistance of multiline varieties. *Annals of Applied Biology* 80, 359-63.

Lerner I.M. (1954) *Genetic Homeostasis.* Oliver & Boyd, Edinburgh & London. 134 pp.

Nelson R.R. (1972) Stabilizing racial populations of plant pathogens by use of resistance genes. *Journal of Environmental Quality* 1, 220-7.

Rajaram S. & Dubin H.J. (1977) Avoiding genetic vulnerability in semi-dwarf wheats. *Annals of the New York Academy of Sciences* 287, 243-54

Rosen H.R. (1949) Oat parentage and procedures for combining resistance to crown rust, including race 45, and Helminthosporium blight. *Phytopathology* 39, 20. (Abstr.)

Simmonds N.W. (1962) Variability in crop plants, its use and conser-vation. *Biological Reviews of the Cambridge Philosophical Society* 37, 422-65.

Vanderplank J.E. (1968) *Disease Resistance in Plants.* Academic Press, New York. 206 pp.

Wolfe M.S. & Barrett J.A. (1980) Can we lead the pathogen astray? *Plant Disease* 64, 148-55.

Race-non-specific disease resistance

J. E. PARLEVLIET
Department of Plant Breeding, Agricultural University,
Wageningen, The Netherlands

INTRODUCTION

In natural ecosystems there are producers (plants), primary consumers
(plants and animals) which satisfy their food requirements directly from
the producers, and higher order consumers. The primary consumers range
from herbivores (caterpillars, sheep) to highly specialized parasites
(powdery mildews, viruses). Hosts have a variety of defence mechanisms
to limit biological damage caused by consumers, and the consumers
in turn evolve adaptations to such defences. In this process of co-
evolution some defence mechanisms are more readily neutralized than
others.

Defence mechanisms can be classified into three groups: avoidance,
resistance and tolerance (Parlevliet, 1980). Avoidance operates before
parasitic contact between host and parasite is established and
decreases the frequency of contact. After parasitic contact has been
established the host can resist the parasite by decreasing its growth,
or tolerate its presence by suffering relatively little damage.

Avoidance is mainly active against animal parasites and includes
such diverse mechanisms as volatile repellents, mimicry, and
morphological features like hairs, thorns and resin ducts (Harper,
1977; Parlevliet, 1977). Resistance is usually of a chemical nature
but little is known about tolerance. It is very difficult to measure
tolerance without confounding it with partial resistance (Parlevliet,
1980).

HOST RESISTANCE

Many important cereal diseases such as the rusts, smuts and mildews,
are caused by very specialized biotrophic fungi. To control these
diseases resistance is commonly used but avoidance and tolerance are
rarely used.

Widespread use of cultivars with very good resistance to these diseases is often followed by the appearance of forms of the pathogen (races) that can overcome these resistances. However, resistance is sometimes maintained and this led Vanderplank (1968) to introduce the concept of vertical (race-specific) and horizontal (race-non-specific) resistance. The former is usually determined by a single gene while the latter is polygenically controlled. Race-specific resistances have been used especially against biotrophic pathogens. When inoculated hosts with this type of resistance often show a hypersensitive reaction, that is infection types (IT) 0, 1 and 2 (Parlevliet, 1979). However, even among host genotypes which give non-hypersensitive, susceptible reactions (IT 3 and 4) large differences in disease severity can occur (Vanderplank, 1968). Such partial resistance often seems to be durable and largely race-non-specific (Parlevliet, 1979).

To distinguish between the two types of resistance it is necessary to measure the resistances of a range of host cultivars against a number of pathogen genotypes (Vanderplank, 1968). If the ranking order for resistance of the cultivars differs with the pathogen race (differential interaction) then vertical resistance (VR) is indicated; if the ranking order is independent of the race then horizontal resistance (HR) is present.

There has been a tendency to group all defence mechanisms that are not race-specific as race-non-specific. This may be confusing, because it implies similarities between them which do not exist. Thus avoidance, as exemplified by the closed flowering habit of some barley cultivars, which excludes the spores of *Ustilago nuda* (loose smut) and *Claviceps purpurea* (ergot), may be unspecific and useful in breeding, but it is a mechanism obviously quite different from true resistance. Even in the various forms of true resistance, including non-host resistance, general resistance and pathogen-specific resistance, the level of specificity varies greatly.

Non-host resistance can be regarded as highly unspecific as it protects the host from nearly all pathogens. However, it may indicate a lack of pathogenicity in the pathogen rather than the presence of resistance in the host. For example barley, usually a non-host for *Sphaerotheca fuliginea* (powdery mildew of melon) can be infected by this pathogen once it has been "induced to susceptibility" by *Erysiphe graminis* f.sp. *hordei* (barley powdery mildew)(Ouchi *et al.*, 1974). Therefore barley is apparently not resistant to *S. fuliginea*, but rather *S. fuliginea* is incapable of establishing a parasitic relationship with barley.

General resistance operates against groups of pathogens. The phytoalexins, produced by many plants following cell damage, are effective against numerous fungi. However, effective pathogens seem able to tolerate or neutralize the phytoalexins produced by their hosts, or to prevent their production. As a consequence phytoalexins do not seem to play a significant role in determining resistance to the most important diseases of crops. It is not invariably true that general resistance is race-non-specific. Avenacin, a glucoside produced by the root tips of oats, protects the roots from various soil pathogens, including *Gaeumannomyces graminis* (take-all). However, a special form of the pathogen, *G. graminis* var. *avenae*, can attack oats because it produces the enzyme avenacinase, which hydrolyses avenacin into products much less toxic to the pathogen (Turner, 1961).

Pathogen-specific resistance operates against one pathogen species and may be either race-specific or race-non-specific. For instance wheat carries many hypersensitive (low IT) resistance genes to *Puccinia graminis*, to *P. recondita*, and to *P. striiformis*. These genes are pathogen-specific and race-specific. Partial resistance, characterized by differences in non-hypersensitive (high IT) reactions is also very pathogen-specific but largely race-non-specific (Parlevliet, 1979). The partial resistances of wheat to *P. graminis* and *P. recondita*, of maize to *P. sorghi* and *P. polysora* and of barley to *P. hordei* and *P. striiformis* are all independent of one another (Parlevliet, 1980). Because partial resistance against one rust does not operate against another, related rust, it cannot be classified as a form of general resistance.

Race-specific resistance is common among biotrophic pathogens, which often have many races, corresponding with many resistance genes in their hosts. Races also occur among necrotrophs (e.g. *Helminthosporium* spp.) and facultative saprophytes (e.g. *Fusarium* spp.) but they are generally far fewer than in biotrophs.

When breeding for disease resistance, general resistance is seldom used. Nearly all resistance currently employed is pathogen-specific and much of it race-specific. Recently race-non-specific resistance has been the subject of increasing attention and it is this type of resistance that is considered in the remaining part of this chapter.

RACE-NON-SPECIFIC RESISTANCE

It is not easy to prove that resistance is race-non-specific because large numbers of host cultivars must be tested against large numbers of pathogen isolates. The characteristics of race-non-specific resistance are exemplified by the barley/*Puccinia hordei* (brown rust) system. Of the six cultivars shown in Table 1, three (Quinn, Sudan and EP75) show race-specific resistance conferred by major genes. The cultivars Berac, Julia and Vada have a susceptible type reaction to all five races in both seedling and adult plant stages but are, nevertheless, relatively resistant in the field.

Differences in partial resistance result from small differences in the components of this resistance accumulated over several pathogen generations (Table 2). The principal effects of partial resistance are on *infection frequency* (IF; the proportion of spores which establish sporulating lesions), *latent period* (LP; the time from infection to first spore production) and *spore production* (SP; the number of spores produced per lesion). Latent period is governed by polygenes and it is likely that IF and SP are similarly determined (Parlevliet, 1978a).

Table 1. Infection types (IT) on six barley cultivars inoculated as seedlings with five races of *Puccinia hordei*. IT 4 is a susceptible reaction, IT 2 (some sporulation) and IT 0 (no sporulation) are resistant (hypersensitive) reactions.

Cultivar	Race 11-1	1-2	18	22	24
Quinn	2	4	0	4	0
Sudan	4	4	4	0	4
EP 75	2	2	0	2	4
Berac	4	4	4	4	4
Julia	4	4	4	4	4
Vada	4	4	4	4	4

Table 2. Partial resistances of five barley cultivars to *Puccinia hordei* race 1-2 (expressed as number of uredosori/tiller and infection frequency (IF), latent period (LP) and spore production (SP), all relative to the most susceptible cultivar) and the estimated number of polygenes governing LP (all measurements relate to adult plants).

Cultivar	Number of uredosori/ tiller	IF	LP	SP	Number of polygenes for LP
L94	2800	100	100	100	0
Sultan	750	65	130	80	3
Volla	115	70	130	110	3
Julia	17	65	160	50	5
Vada	1	40	190	50	6

When the race-specificity of this partial resistance was tested by exposing three resistant cultivars to five different brown rust races (Table 3) the low to moderate disease severity was by and large consistent with race-non-specific resistance. Cultivar Vada was the most resistant against all five races and cultivar Berac usually the least resistant. However, there was one small but significant

exception; cv. Julia always had less brown rust than cv. Berac,
except when exposed to race 18 against which cv. Julia was the more
susceptible. This differential interaction was due to a difference in
LP. Cultivars Berac and Julia gave a similar LP for four of the five
races but race 18 had a LP 1 to 1.5 days shorter on cv. Julia than on
cv. Berac. This corresponds well with the estimated effect of one
polygene. The genetic and epidemiological data suggest that one of
the polygenes for LP in cv. Julia has been "broken" by race 18
(Parlevliet, 1978b), although the size of this race-specific effect
is too small to be used to readily identify races. Similar small,
race-specific differences in partial resistance have also been reported
for the potato/*Phytophthora infestans*, barley/*Rhynchosporium secalis*,
wheat/*Septoria tritici* and wheat/*Puccinia recondita* systems (reviewed
by Parlevliet, 1979).

Table 3. Disease severity (% sporulating leaf area) on three
barley cultivars infected by five races of *Puccinia hordei*, just
prior to maturity.

Cultivar	Race 11-1	18	1-2	22	24
Berac	8.1	6.7	3.1	5.0	0.9
Julia	4.5	12.1	1.8	1.1	0.6
Vada	0.8	0.5	0.6	0.2	0.1

Durability

Race-specificity is generally considered as evidence for the probable
breakdown or erosion of resistance but this is not necessarily true.
Race-specific resistance against biotrophic pathogens, generally of a
major gene, hypersensitive type, is highly unstable, but major gene
resistance against non-biotrophs, although also race-specific, is
often considerably more durable. The durability of partial resistance
seems even greater. Many Western European barley cultivars have good
partial resistance to brown rust even though the only selection for
resistance has been to remove the most susceptible lines from breeding
programmes (Parlevliet, 1980). Some of the polygenes for partial
resistance occur with a high frequency (Parlevliet, 1978a) and must
have been present for some time as they have not been consciously
introduced or selected recently. As they are still effective they must
be very durable despite being sometimes race-specific.

Recognition of race-non-specific resistance

Race-specificity and lack of durability are predominantly problems with resistances to biotrophic pathogens, so it is especially important to distinguish race-specific from partial resistance to them. The race-specific resistances are generally characterized by a hypersensiti⁀ or low IT reaction, which may result in no sporulation (IT 0, complete resistance) or in restricted sporulation (IT 1 or 2, incomplete resistance). To distinguish partial resistance from complete resistance is fairly easy. To distinguish it from incomplete resistance, however, is far more difficult, as both allow some disease. The breeder is therefore advised to select plants or lines with a susceptible IT but nevertheless having relatively little disease, in order to increase his chances of selecting partial resistance.

Genetics of race-non-specific resistance

In the relationships between hosts and biotrophic pathogens, resistance and virulence genes often operate on a gene-for-gene basis (Day, 1974). When effects of resistance and virulence genes are large, it is easy to differentiate host cultivars carrying different resistance genes and pathogen races carrying different virulence genes. In the hypothetical example in Table 4 the differential interaction is between the host cultivars $R_1R_1r_2r_2$ and $r_1r_1R_2R_2$ and the pathogen races aB and Ab and is a typical case of vertical resistance.

Table 4. Hypothetical example illustrating relative disease severities where a gene-for-gene relationship operates. It is assumed that the host is homozygous diploid and the pathogen haploid. R_1 and R_2 are resistance alleles, a and b are virulence alleles at two loci that neutralize the effects of R_1 and R_2 respectively.

Host genotype	Pathogen genotype			
	AB	aB	Ab	ab
$r_1r_1r_2r_2$	100	100	100	100
$R_1R_1r_2r_2$	50	100	50	100
$r_1r_1R_2R_2$	50	50	100	100
$R_1R_1R_2R_2$	0	50	50	100

Absence of a differential interaction, usually taken to indicate horizontal or race-non-specific resistance, may arise in two ways (Parlevliet & Zadoks, 1977). If there is no gene-for-gene relationship, no differential interaction is possible. In these circumstances the disease severities of the four host-pathogen combinations giving the differential interaction in Table 4 will all be 100% as both of the pathogen genes a and b will neutralize R_1 and R_2. However, a system in which there is no gene-for-gene relationship and no race-specificity is likely to be even less stable than one in which they do operate (Parlevliet & Zadoks, 1977). Nevertheless, a gene-for-gene relationship is not incompatible with race-non-specific resistance. When the gene effects are large, the differential interactions, which are equal to the gene effects (Table 4), are also large and easily detected and one speaks of race-specifity. If the gene number is larger but the effects of individual genes smaller, the differential interactions are smaller and more difficult to recognize. When the sizes of differential interactions are of the same order of magnitude as the experimental error they cannot be identified, races cannot be recognized and one speaks of race-non-specific resistance.

Accumulating evidence suggests that most, if not all, resistance genes operate on a gene-for-gene basis with virulence genes in the pathogen. The apparent degree of race-specificity of resistance varies in a continuous way from typical VR, when the genes involved have large effects, to near HR in the case of polygenic systems. Partial resistance derives its race-non-specific character from relatively numerous resistance genes that operate on a gene-for-gene basis with corresponding virulence genes in the pathogen but all having only small effects. For example Zadoks (1972) tried to relate the wheat/ *Puccinia striiformis* system to Vanderplank's concept of VR and HR but he considered the clear-cut division between the two to be unrealistic as he observed a continuum between instances of clear VR and near HR.

We still need to explain why low IT resistance to biotrophic pathogens is more unstable than partial resistance. Parlevliet (1980) suggested that there are two types of pathogen-specific resistance gene, one relatively stable and the other very unstable. The former consists of resistance genes, that are assumed to operate within the basic resistance-pathogenicity system and confer partial resistance. The latter genes are thought to act within an incompatibility system giving low IT reactions, superimposed upon this basic system, and are used by the biotrophic pathogen as a genetic feed-back mechanism to regulate their coexistence with the host at intermediate levels of pathogenicity. Biotrophs need such a feed-back mechanism as too great a pathogenicity puts the pathogen population indirectly at risk by endangering its host, while insufficient pathogenicity endangers it directly. The introduction of low IT genes into crop cultivars activates these feed-back mechanisms stimulating a rapid shift from avirulence to virulence.

REFERENCES

Day P.R. (1974) *Genetics of host-parasite interaction*. W.H. Freeman, San Francisco, 238 pp.
Harper J.L. (1977) *Population biology of plants*. Academic Press, London, 892 pp.
Ouchi S., Oku H., Hibino C. & Akiyama I. (1974) Induction of accessibility to a non pathogen by preliminary inoculation with a pathogen. *Phytopathologische Zeitschrift* 79, 142-54.

Parlevliet J.E. (1977)   Plant pathosystems: an attempt to elucidate horizontal resistance. *Euphytica* 26, 553-6.

Parlevliet J.E. (1978a)   Further evidence of polygenic inheritance of partial resistance in barley to leaf rust, *Puccinia hordei*. *Euphytica* 27, 369-79.

Parlevliet J.E. (1978b)   Race-specific aspects of polygenic resistance of barley to leaf rust, *Puccinia hordei*. *Netherlands Journal of Plant Pathology* 84, 121-6.

Parlevliet J.E. (1979)   Components of resistance that reduce the rate of epidemic development. *Annual Review of Phytopathology* 17, 203-22.

Parlevliet J.E. (1980)   Disease resistance in plants and its consequence for breeding. *Proceedings Plant Breeding Symposium II* (Ed. by K.J. Frey), Iowa State University, Ames. In press.

Parlevliet J.E. & Zadoks J.C. (1977)   The integrated concept of disease resistance; a new view including horizontal and vertical resistance in plants. *Euphytica* 26, 5-21.

Turner E.M.C. (1961)   An enzymic basis for pathogenic specificity in *Ophiobolus graminis*. *Journal of Experimental Botany* 12, 169-75.

Vanderplank J.E. (1968)   *Disease resistance in plants*. Academic Press, London, 206 pp.

Zadoks J.C. (1972)   Modern concepts of disease resistance in cereals. In *The way ahead in plant breeding* (Ed. by F.G.H. Lupton, G. Jenkins & R. Johnson), pp. 89-98.   Plant Breeding Institute, Cambridge.

Durable disease resistance

R. JOHNSON

Plant Breeding Institute, Trumpington, Cambridge

INTRODUCTION

This chapter discusses the characteristics of durable disease resistance. It is restricted to host resistance and does not refer to tolerance (Parlevliet, this volume). A method for breeding cultivars with
disease resistance that should persist during widespread use is described. The method can be applied using existing techniques of
selection for disease resistance.

THE CONCEPT OF DURABLE RESISTANCE

The plant breeder aims to introduce into cultivars disease resistance
that will give useful disease control for long enough to ensure that
the commercial life of a cultivar is not curtailed by a breakdown in
disease resistance. Prolonged resistance may result if a pathogen
is unable to respond to host resistance with matching cultivar-specific
pathogenicity as, apparently, with some necrotrophic parasites. For
example, resistance in the wheat cultivar Cappelle-Desprez to *Pseudocercosporella herpotrichoides* controlled eyespot for more than 20 years
when the cultivar was extensively grown in Western Europe (Scott *et al.*,
1979). By contrast, resistance to many other pathogens, especially
the biotrophic rusts and powdery mildews, is often highly race-specific.
Failures of such resistance result when pathogen races with specific
matching pathogenicity occur but this may cause little economic loss
if the pathogen spreads slowly or the crop is genetically diverse.
Losses may be great, however, with diseases such as the rusts and
powdery mildews because the pathogens multiply rapidly and spread by
easily-dispersed airborne spores on crops that are often genetically
uniform and occupy large areas. Some cultivars, however, remain
resistant to pathogens that have highly developed cultivar-specific
pathogenicity even though they are extensively cultivated in environments favourable to disease. They can then be described as possessing
durable resistance (Johnson & Law, 1973, 1975; Johnson, 1979). There
are several examples of durable resistance to rust diseases of wheat
(Johnson, 1978; Line, 1978) and to diseases of other crops (Eenink,

1977).

It is necessary to distinguish the description of disease resistance as durable from some other terms used to describe resistance that might be expected to be durable. If resistance were equally effective against all races of a pathogen it could be described as race-non-specific or, according to Vanderplank (1963, 1978), as horizontal. Clearly such resistance might be expected to be durable and it is for this reason that plant breeders have been interested in it (Howard *et al.*, 1970). It is impossible, however, to prove that resistance is horizontal because to do so requires a demonstration that the pathogen is not capable of evolution towards increased pathogenicity specific for that resistance. Horizontal resistance is therefore a theoretical concept. This allows some to believe that all plants possess horizontal resistance (Robinson, 1976) while others hold that all resistance is race-specific and operates through a gene-for-gene interaction with specific pathogenicity in the pathogen (Parlevliet, this volume); Ellingboe (1975) suggested that horizontal resistance is an artefact and is merely resistance that has not yet been shown to be race-specific. By contrast the existence of durable resistance is an observed fact in certain cultivars.

RECOGNITION OF DURABLE RESISTANCE

Certain mechanisms that confer resistance are likely to provide durable protection. Barley cultivars with flowers that do not open are unlikely to be infected by the loose smut fungus (*Ustilago nuda*) (Macer, 1964). Although the stigmata may be susceptible, infection is avoided (Parlevliet, this volume). An improbably large change in the biology of the pathogen would be required to overcome this barrier which is likely to provide both durable and race-non-specific resistance.

Usually, durable resistance cannot be so easily identified. In particular it must be noted that resistance that is incomplete need not be durable. This must be emphasized because the erroneous idea that incomplete or partial resistance is race-non-specific or horizontal has gained wide acceptance. This arose from Vanderplank's observation that potato genes conferring complete resistance to *Phytophthora infestans* were race-specific whereas intermediate levels of resistance in old cultivars were apparently not and remained effective (J.E. Vanderplank, personal communication). Such a distinction between complete and incomplete resistance does not apply to some other host-pathogen combinations. There are numerous examples of wheat cultivars where incomplete resistance to yellow rust is only effective against some isolates of *Puccinia striiformis* (Manners, 1950; Zadoks, 1961; Johnson & Taylor, 1972; Johnson & Bowyer, 1974; Priestley, 1978). Between 1970 and 1977 five wheat cultivars were tested at Cambridge. Over this period four of them showed a marked reduction in resistance to *P. strii-formis* due to the evolution of two new races, one with increased pathogenicity specific for cultivars Maris Huntsman and Maris Nimrod, the other for cultivars Maris Bilbo and Hobbit (Table 1) (Johnson &

Taylor, 1980). The resistance of cv. Cappelle-Desprez did not change. Although the other four cultivars all possessed incomplete resistance in 1970 their resistance contained race-specific components of various sizes. In the case of cv. Maris Huntsman, its resistance in 1977 was still adequate for commercial exploitation of the cultivar but that of cultivars Maris Nimrod and Maris Bilbo was clearly too low.

Table 1. Severity of yellow rust (% leaf area affected) on five artificially inoculated wheat cultivars grown in field plots in 1970 and 1977.

Cultivar	July 1970	July 1977	Difference
Cappelle Desprez	42	43	+ 1
Maris Nimrod	45	77	+32
Maris Huntsman	20	40	+20
Maris Bilbo	20	88	+68
Hobbit	6	67	+61
SED	4.1	4.6	

No single mechanism or genetic model adequately explains durable resistance. The only reliable criterion for its recognition is, therefore, that the resistance remains effective after extensive and prolonged use (Johnson, 1978). It is not necessary to claim that such resistance is race-non-specific, nor that it will be permanent, although it may be hoped that the longer the resistance has been effective the longer it will remain so.

Limitations of testing at many locations

All new cultivars must be tested in a range of environments and this is usually achieved by growing them at different sites. It is commonly assumed that such tests will expose cultivars to more pathogen races than would be encountered at a single site. Cultivars whose resistance is stable over many sites and years are often selected in preference to others that are resistant at some sites and susceptible at others. However, the distribution of races between sites is frequently not known and the trials cannot provide a powerful test of the durability of resistance because the total area occupied by a cultivar in such trials is small compared with the area that would be occupied if it were adopted for commercial use. The chances of detecting rare races with specific pathogenicity for a particular cultivar are therefore small and the test of durability is weak (Johnson, 1978).

Differences in disease severity on a cultivar at different sites can also be due to differences in the environment between sites and may not indicate race-specificity. It is possible for a cultivar to have resistance that is adequate, and may be durable, in one environment but inadequate in another environment more favourable to disease. For example, resistance of the wheat cultivar Cappelle-Desprez to yellow rust was both adequate and durable in Britain but inadequate in Sweden (A. Hagberg, personal communication).

GENETICS OF DURABLE RESISTANCE

Many hypotheses have been proposed to describe the genetic control of race-non-specific resistance. It has been suggested that genes conferring race-specific resistance are different from those conferring race-non-specific resistance (Robinson, 1976). Another view is that both types of resistance are controlled by the same genes whose individual and combined effects differ (Nelson, 1978). Parlevliet & Zadoks (1977) implied that durable resistance would be controlled by numerous genes each having a small effect. They proposed a theoretical model to show that such resistance would show greater stability in its expression if the genes for resistance interacted in a gene-for-gene relationship with corresponding genes in the pathogen, than if the resistance genes acted additively and race-non-specifically. Heather *et al.* (1980) suggested, however, that the durability of the resistance of the interactive model would depend on the ability of the pathogen to accumulate alleles with specific pathogenicity for the individual resistance genes.

Several authors have concluded from their data that incomplete, general or long-lasting resistance to rust pathogens, presumably of the type thought to be durable, is controlled by small numbers of genes with additive action (Sharp & Volin, 1970; Skovmand & Wilcoxson, 1975; Knott, 1977). This has led to the suggestion that such resistance can be accumulated by intercrossing susceptible cultivars and selecting for transgressive levels of resistance (Pope, 1968; Robinson, 1976; Sharp, 1976). Robinson (1976) suggested that, to avoid the selection of race-specific resistance genes in the progeny, all the parents should be very susceptible to a single race of the pathogen that is used to assess resistance in the progeny. Clearly, if the parents carry race-specific resistance genes and are each susceptible to a different race of the pathogen, apparently transgressive segregation for resistance towards the available races may be due to the recombination of race-specific genes. This resistance may not remain effective if pathogen races develop with corresponding combinations for specific pathogenicity. If the interactive model of Parlevliet & Zadoks (1977) is correct, however, even transgressive segregation from parents that are susceptible to the same races may be controlled by genes that are race-specific. The durability of resistance might then depend on obtaining particular gene combinations or on accumulating large numbers of genes, which is impracticable in most breeding programmes.

It is possible that durable resistance may be controlled by genetic systems that are not exceptionally difficult to manipulate in breeding programmes. Johnson & Law (1973, 1975) showed that a single chromosome, designated 5BS-7BS, controlled much of the adult plant resistance in four genetically related cultivars with a history of durable resistance to yellow rust. The chromosome carried at least three factors for resistance (Law et al., 1978) and it was considered that it might also be responsible for the durability of resistance. However, in one of the cultivars, Cappelle-Desprez, monosomic analysis showed that several chromosomes in addition to 5BS-7BS affected resistance. To test the hypothesis that 5BS-7BS controls durability of resistance it will be transferred to other cultivars to see whether it confers enhanced durability in them (Johnson & Law, 1975; Law et al., 1978). Early evidence from this work indicates, however, that the expression of resistance conferred by this chromosome and so, possibly, the durability, is dependent on the genetic background into which it is introduced (J. Bingham, personal communication).

Origin of cultivars with durable resistance

The pedigrees of wheat cultivars with durable resistance to yellow rust suggest that it is heritable. Several cultivars probably derived their durable resistance from the cultivar Wilhelmina (Mesdag, 1975), while durable resistance to yellow rust in Cappelle-Desprez was probably inherited from Vilmorin 27 (Johnson, 1978). Nevertheless durable resistance has probably originated by chance from diverse sources. The resistant Dutch cultivar Manella was derived from a cross between cultivars Heines VII and Alba, both of which are very susceptible to different races of *P. striiformis*, although no single race that attacks both cultivars to the same extent has been reported (Stubbs et al., 1974). While it is possible that the resistance of cv. Manella is due to a combination of race-specific components derived from each cultivar this seems unlikely because the pathogen would be expected to have recombined the necessary specific factors for pathogenicity to overcome its resistance. In practice, however, the resistance of cv. Manella has proved durable (Stubbs et al., 1974).

   Some resistance to rusts and powdery mildews has been derived from related species and genera. Often it has had no greater durability than the race-specific resistance already available within the species (Scott et al., 1979). However, an apparently durable type of resistance to *P. graminis* in the wheat cultivar Hope (Hare & McIntosh, 1979) was derived from a cross between the tetraploid wheat Yaroslav Emmer and a hexaploid cultivar Marquis (McFadden, 1930). A gene *Sr26*, for resistance to *P. graminis*, derived from *Agropyron elongatum* has also apparently remained effective for a long period in Australia (R.A. McIntosh, personal communication). Durable resistance to powdery mildew in cv. Maris Huntsman winter wheat (Scott et al., 1979) was probably derived from *Triticum timopheevi*.

   Although the genetic control of durable resistance to rusts and powdery mildews usually seems to be more complex than the genetic control of recognizably race-specific resistance that lacks durability, there is as yet too little evidence to support a definitive genetic model. This is partly due to the difficulty of recognizing the durability of resistance, as distinct from the resistance itself, and partly because the degree of genetic complexity controlling durable resistance appears to differ in different cases. Eenink (1977) listed several examples of durable resistance to various diseases, including maize and oats resistant to *Helminthosporium* spp., that were apparently under the genetic control of a single or of only a few genes.

PRACTICAL BREEDING FOR DURABLE RESISTANCE

Programmes based on available cultivars with durable resistance

Cultivars with recognized durable resistance would seem to be the obvious donors of durable resistance in breeding programmes. In exploiting such resistance in practical breeding programmes the production of effective recombinations of recognized race-specific genes which are unlikely to be durable should be avoided because they will complicate testing for durable resistance. The simplest way to avoid this problem is to ensure that all parents in the crossing programme are susceptible to the same race as that to which the donor of durable resistance shows its maximum known susceptibility and that this race is used to test the resistance of the progeny (Johnson, 1978). Clearly a detailed knowledge of host-specific pathogenicity in the pathogen, on both adult plants and seedlings, is necessary. This procedure also imposes restrictions on the choice of parents for crossing. These restrictions are not always acceptable to the breeder who must select for many characters in addition to disease resistance. Sometimes, therefore, effective combinations of race-specific genes will inevitably occur but selection against these may still allow resistance from a durably resistant parent to be detected (Johnson, 1978).

Avoiding recombination of specific resistance genes that will hinder the selection of resistance from durably resistant parents is the most difficult part of transferring durable resistance to new cultivars. It is still easier, however, than attempting to accumulate many resistance genes of small effect against a single disease.

Breeding for durable resistance when sources of it have not been recognized

The breeder may wish to breed for durable resistance to diseases before sources of such resistance have been identified and often when little is known about host-specificity in the pathogens. It is unlikely that a breeder would be willing to start breeding for resistance to disease by crossing very susceptible cultivars, as suggested by Robinson (1976), and in any case there is little evidence that such a policy would be generally successful. The most practical policy, therefore, is to seek sources of resistance, regardless of type, to use in the breeding programme and, at the same time, to assess the potential for host-specific pathogenicity in the pathogen. When resistant cultivars are released they should be closely monitored and, if possible, cultivars with resistance derived from other sources should be available to replace any whose resistance fails. Diversity in the use of cultivars should also be encouraged (Priestley, this volume). Even if some resistance lacks durability, cultivars with durable resistance may eventually be identified for use in further breeding programmes. If not, it may be necessary to resort to the use of multilines or, preferably, cultivar mixtures for disease control (Wolfe *et al.*, this volume), or perhaps to intercross susceptible lines and look for transgressive resistance as suggested by Robinson (1976).

Selection criteria for durable resistance in a breeding programme

Most breeding programmes involve screening large numbers of plants for many characters and there is an obvious advantage in using the simplest possible criteria for selection. In selecting for resistance to foliar pathogens, such as rusts and powdery mildews, visual estimation of the proportion of leaf tissue affected by the disease is the commonest and quickest method. It has proved sufficiently accurate for selecting rust resistant wheats at the Plant Breeding Institute provided only that there is sufficient inoculum in the plots (Lupton & Johnson, 1970).

    Although latent period and sporulation can be correlated with disease severity in field tests (Parlevliet & van Ommeren, 1975; Parlevliet, this volume) they are less useful than visual estimation as selection criteria because their measurement requires more labour and this would limit the amount of material that could be screened. Measuring such a component of resistance might by justified if it was known to control durability, if it was quicker and more accurate than the visual method or if it could be done in the absence of the pathogen.

CONCLUSIONS

The method of identifying and breeding for durable disease resistance described here may be the simplest that can be devised with present knowledge although eventually we may learn how durable resistance is controlled so that it can be better exploited. If the method is successful it should provide cultivars that can be tested and grown using methods already employed for existing genetically uniform cultivars. By contrast, the use of cultivar mixtures might require modifications in testing and cultural techniques. Moderate levels of durable resistance could enhance the value of other disease control methods and decrease the risk of sudden failure of resistance due to increased cultivar-specific pathogenicity.

REFERENCES

Eenink A.H. (1977) Genetics of host-parasite relationships and the stability of resistance. *Induced Mutations Against Plant Diseases. Proceedings of an International Atomic Energy Symposium, Vienna* 1977, pp. 47-56.

Ellingboe A.H. (1975) Horizontal resistance: an artefact of experimental procedure? *Australian Plant Pathology Society Newsletter* 4, 44-6.

Hare R.A. & McIntosh R.A. (1979) Genetic and cytogenetic studies of durable adult plant resistance in 'Hope' and related cultivars to wheat rust. *Zeitschrift für Pflanzenzüchtung* 83, 350-67.

Heather W.A., Sharma J.K. & Miller A.G. (1980) Physiologic specialisation in *Melampsora larici-populina* Kleb. on clones of poplar demonstrating partial resistance to leaf rust. *Australian Forestry Research* 10, 125-31.

Howard H.W., Johnson R., Russell G.E. & Wolfe M.S. (1970) Problems in breeding for resistance to diseases and pests. *Report of the Plant Breeding Institute for 1969*, pp. 6-36.

Johnson R. (1978) Practical breeding for durable resistance to rust diseases in self-pollinating cereals. *Euphytica* 27, 529-40.

Johnson R. (1979) The concept of durable resistance. *Phytopathology* 69, 198-9.

Johnson R. & Bowyer D.E. (1974) A rapid method for measuring production of yellow rust spores on single seedlings to assess differential interactions of wheat cultivars with *Puccinia striiform. Annals of Applied Biology* 77, 251-8.

Johnson R. & Law C.N. (1973) Cytogenetic studies on the resistance of the wheat variety Bersée to *Puccinia striiformis. Cereal Rusts Bulletin* 1, 38-43.

Johnson R. & Law C.N. (1975) Genetic control of durable resistance to yellow rust (*Puccinia striiformis*) in the wheat cultivar Hybride de Bersée. *Annals of Applied Biology* 81, 385-91.

Johnson R. & Taylor A.J. (1972) Isolates of *Puccinia striiformis* collected in England from the wheat varieties Maris Beacon and Joss Cambier. *Nature, London* 238, 105-6.

Johnson R. & Taylor A.J. (1980) Pathogenic variation in *Puccinia striiformis* in relation to the durability of yellow rust resistance in wheat. *Annals of Applied Biology* 94, 283-6.

Knott D.R. (1977) Studies on general resistance to stem rust. *Induced Mutations Against Plant Diseases. Proceedings of an International Atomic Energy Agency Symposium, Vienna* 1977, pp. 81-6.

Law C.N., Gaines R.C., Johnson R. & Worland A.J.(1978) The application of aneuploid techniques to a study of stripe rust resistance in wheat *Proceedings of the Fifth International Wheat Genetics Symposium, New Delhi* 1978, pp. 427-46.

Line R.F. (1978) Pathogenicity of *Puccinia striiformis* and resistance to stripe rust in the United States. *Abstracts of papers, Third International Congress of Plant Pathology, Munich* 1978, p.305.

Lupton F.G.H. & Johnson R. (1970) Breeding for mature-plant resistance to yellow rust in wheat. *Annals of Applied Biology* 66, 137-43.

Macer R.C.F. (1964) Developments in cereal pathology. *Report of the Plant Breeding Institute for 1962-63*, pp. 5-33.

McFadden E.S. (1930) A successful transfer of emmer characters to *vulgare* wheat. *Journal of the American Society of Agronomy* 22,

1020-34.

Manners J.G. (1950)   Studies on the physiologic specialization of yellow rust (*Puccinia glumarum* (Schm.) Erikss. & Henn.) in Great Britain. *Annals of Applied Biology* 37, 187-214.

Mesdag J. (1975)   Sortenwechsel bei Winterweizen in den Niederlanden. *Bericht über die Arbeitstatung 1975 in der 'Arbeitsgemeinschaft der Saatzuchtleiter'*, pp. 79-90.

Nelson R.R. (1978)   Genetics of horizontal resistance. *Annual Review of Phytopathology* 16, 359-78.

Parlevliet J.E. & van Ommeren A. (1975)   Partial resistance of barley to leaf rust, *Puccinia hordei*. II. Relationship between field trials, microplot tests and latent period. *Euphytica* 24, 293-303.

Parlevliet J.E. & Zadoks J.C. (1977)   The integrated concept of disease resistance; a new view including horizontal and vertical resistance in plants. *Euphytica* 26, 5-21.

Pope W.K. (1968)   Interaction of minor genes for resistance to stripe rust of wheat. *Proceedings of the Third International Wheat Genetics Symposium, Canberra 1968*, pp.251-7.

Priestly R.H. (1978)   Detection of increased virulence in populations of wheat yellow rust. *Plant Disease Epidemiology* (Ed. by P.R. Scott & A. Bainbridge), pp. 63-70.   Blackwell Scientific Publications, Oxford.

Robinson R.A. (1976) *Plant Pathosystems*.   Springer-Verlag, Berlin. 184 pp.

Scott P.R., Johnson R., Wolfe M.S., Lowe H.J.B. & Bennett F.G.A. (1979) Host specificity in cereal parasites in relation to their control. *Report of the Plant Breeding Institute for 1978*, pp. 27-62.

Sharp E.L. (1976)   Broad based resistance to stripe rust in wheat. *Proceedings of the Fourth European and Mediterranean Cereal Rusts Conference, Interlaken 1976*, pp.159-61.

Sharp E.L. & Volin R.B. (1970)   Additive genes in wheat conditioning resistance to stripe rust. *Phytopathology* 60, 1146-7.

Skovmand B. & Wilcoxson R.D. (1975)   Inheritance of slow rusting in spring wheat to stem rust. *Annual Proceedings of the American Phytopathology Society* 2, 90-1.

Stubbs R.W., Fuchs E., Vecht H. & Bassit E.J.W. (1974)   The international survey of factors of virulence of *Puccinia striiformis* Westend. in 1969, 1970 and 1971. *Stichting Nederlands Graan-Centrum, Technisch Bericht* 21, 16.

Vanderplank J.E. (1963) *Plant Diseases: Epidemics and Control* Academic Press, New York.   349 pp.

Vanderplank J.E. (1978) *Genetic and Molecular Basis of Plant Pathogenesis*. Springer-Verlag, Berlin. 167 pp.

Zadoks J.C. (1961)   Yellow rust of wheat: studies in epidemiology and physiologic specialization. *Tijdschrift over Plantenziekten* 67, 69-256.

Choice and deployment of resistant cultivars for cereal disease control

R. H. PRIESTLEY

National Institute of Agricultural Botany,
Huntingdon Road, Cambridge

INTRODUCTION

The aim of any cereal disease control strategy is to manage the local
pathogens so that yield losses are minimized. It is not necessary to
eradicate the pathogens but merely to keep their incidence below that at
which significant losses occur. This chapter is concerned with some of
the factors affecting the choice of resistant cultivars by farmers and
discusses how those chosen can be deployed most effectively to control
diseases.

CHOICE OF CULTIVARS

Relative importance of yield, quality and resistance

Cereal cultivars differ widely in many characters apart from disease
resistance. Although plant breeders have not yet been able to produce
cultivars combining high levels of all desirable characters, those that
are commercially successful usually possess at least one of them to a
high degree. Potential yield is usually the most important character,
although quality can be of overriding importance in crops grown for
milling or malting.

 Disease resistance is important only if diseases decrease yield.
When a disease has been shown to be damaging, resistance to it is
especially important where the choice is between cultivars of similar
potential yield. Once the relationship between disease and yield loss
has been established, the merits of a relatively low yielding but
resistant cultivar can be compared with those of a potentially high
yielding but susceptible cultivar.

Deciding which diseases need to be controlled

The first step in choosing which cultivars to grow on a particular farm
is to determine those diseases that are likely to need controlling.
Surveys of commercial crops of spring barley and winter wheat (King,

1972, 1977a, 1977b) have clearly shown powdery mildew (*Erysiphe graminis* to be important nationally. However, diseases which are relatively unimportant nationally may be limiting in particular localities where the climate favours them. Unfortunately the areas in which different diseases become severe can also change from year to year (King, 1972, 1977b), presumably reflecting variations in local climate. Because diseases are so variable in their distribution from year to year, disease frequency, expressed as the proportion of years that a critical disease severity occurs in a particular area, gives the best indication of the need to control them.

Cultivar trials such as those done by the National Institute of Agricultural Botany (NIAB) are good sources of information as they are usually done at the same sites for many years and are routinely assessed for diseases. Data from NIAB trials at 14 sites during the 20 years 1957 to 1976 have been used to produce disease frequency maps for various rusts (Priestley, 1978), powdery mildew (Priestley & Bayles, 1979a), rhynchosporium and septoria (Priestley & Bayles, 1979b) in England and Wales. More recently, the positions of lines linking areas of equal disease frequency (isopaths) have been mapped using an unpublished model that assumes an exponential relationship between disease frequency and distance between neighbouring sites. As an example the distribution of *Rhynchosporium secalis* determined in this way is shown in Fig 1. More sites would clearly improve the accuracy of locating the isopaths but simplified versions of the existing maps (Anon., 1979a) are being used by farmers to determine the likelihood of various diseases occurring in their areas.

Resistance to disease

The risk of serious disease occurring on a cultivar is largely determined by its genetic resistance to the most prevalent pathogens in the areas in which it is grown. Many countries publish ratings, on a 0-9 scale, indicating the resistances of locally available cultivars to various pathogens. In England and Wales, this information is published annually by NIAB in Farmers' Leaflet No 8 - 'Recommended varieties of cereals'. The resistance ratings are calculated from measurements of natural infection in cultivar trials at many locations and from results of field experiments in which adult plants are inoculated with pathogen isolates representing the range of pathogen variation found by the United Kingdom Cereal Pathogen Virulence Survey (Anon., 1979b).

Before being recommended for use in England and Wales, cultivars must be shown to have a minimum level of resistance to the most important pathogens (Doodson, 1976). The aim is to dissuade breeders from producing, and farmers from growing, cultivars that are very susceptible to particular pathogens.

Resistance and response to fungicides

Genetic resistance is not the only means of controlling diseases and the likely yield response to fungicides can be an important consideration

Figure 1. Isopaths of *Rhynchosporium secalis* determined from
cultivar trial data collected at 14 sites during the 20 **years** 1957
to 1976.

● = trial site
Figures are numbers of years out of 20 when at least 10% of the leaf
area was infected.

when choosing cultivars. About 50% of spring barley **crops** in England
and Wales are currently treated with fungicides, many of them **prophy**-
lactically for the control of mildew. Spring barley trials **which**
measured the effects of fungicides applied to control mildew **showed** that
cultivars differ in their yield response (Rowe & Doodson, 1976) and this
can affect the rank order of yield of cultivars, particularly if some
are very susceptible. However, the trials also showed that the 0-9
resistance ratings are a good guide to the average yield **response** that
can be expected when disease is controlled by fungicides. **Farmers may**
eventually be able to estimate the relative yields **of cultivars when**
treated with fungicides by adding to the published yield values suitable
increments calculated from their resistance ratings.

DEPLOYMENT OF CULTIVARS ON THE FARM

Epidemics are most likely to occur if a single cultivar is grown on a large area in successive years. Analyses of spore populations of *E. graminis* and *Puccinia striiformis* have shown that inoculum generated by a cultivar possessing a specific resistance is largely non-virulent on cultivars possessing other specific resistances (Wolfe & Wright, 1978; Priestley, 1979). Thus the spread of disease will be hindered if neighbouring fields are sown with cultivars possessing different specific resistances. To exploit this effect, diversification schemes have been devised to assist farmers in choosing cultivars to minimize the risk of epidemics of *P. striiformis* on winter wheat and *E. graminis* on spring barley (Priestley & Wolfe, 1977). A simplified version of the 1979 scheme for spring barley (Anon., 1979a) is shown as an example in Tables 1 and 2.

Table 1. Diversification groups (DG) of spring barley cultivars, based on specific resistances (R) effective against *Erysiphe graminis*

DG 0	Armelle (R 0)
DG 1	Magnum (R 4+?); Simon (R 8)
DG 2	Midas (R 3)
DG 3	Georgie (R 2+4); Sundance (R 2+4)
DG 4	Goldmarker (R 3+4); Jupiter (R 3+4)
DG 6	Ark Royal (R 6); Mazurka (R 2+6); Wing (R 6)
DG 7	Tyra (R 7)

Cultivars are placed in diversification groups (DG) on the basis of their specific resistances (R factors) identified using methods described by Priestley & Wolfe (1977). The method used to classify resistances against *P. striiformis* has recently been modified to use dendrograms produced by multivariate analysis to compare new cultivars with control cultivars possessing known resistances (Priestley & Byford, 1979). In the schemes for both wheat yellow rust and barley mildew, cultivars possessing either no specific resistances or those that have been rendered ineffective by universal virulence are placed in DG 0. Cultivars possessing resistances that are still completely effective, or nearly so, are placed in DG 1. Cultivars possessing resistances rendered partly ineffective by the development of specific virulence in the pathogen are grouped with other cultivars possessing similar specific resistances. In the current spring barley scheme, cultivars are classified into seven groups (Table 1). The relationships between the phenotypic resistances R 2 to R 8 and known resistance genes have been described by Wolfe & Slater (1979).

Farmers can choose cultivars to grow in neighbouring fields by using a selection matrix (Table 2). In general, cultivars from within the

Table 2. Matrix for selecting spring barley cultivars from different diversification groups (DG) to reduce spread of *Erysiphe graminis*

Chosen DG	DGs that can reduce spread of *E. graminis* when grown with chosen DG						
	DG 1	DG 2	DG 3	DG 4	DG 5	DG 6	DG 7
DG 1	+	+	+	+	+	+	+
DG 2	+	•	+	•	+	+	+
DG 3	+	+	•	•	+	+	+
DG 4	+	•	•	•	+	+	+
DG 5	+	+	+	+	•	+	+
DG 6	+	+	+	+	+	•	+
DG 7	+	+	+	+	+	+	•

+ signs indicate those combinations of DGs that can reduce spread of *E. graminis*

same group should not be grown in close proximity to one another. Cultivars in DG 1 are exceptional because their resistance is still completely effective, or nearly so, so that they are unlikely to become infected and generate much inoculum. They can, therefore, be grown near any other cultivar including others in DG 1.

The schemes are revised annually using the results of the United Kingdom Cereal Pathogen Virulence Survey. Comprehensive schemes for the whole of the United Kingdom are produced (Anon., 1979b) and then modified for local use in England and Wales (Anon., 1979a, 1979c) and Scotland (Anon., 1979d). A survey of farmers in 1979 showed that 56% of those questioned used the wheat yellow rust scheme and 28% used the spring barley mildew scheme. More intimate diversity can be achieved by mixing seed of different cultivars for sowing in the same field (Wolfe, 1978; Wolfe *et al.*, this volume) and the diversification schemes can be used to choose the appropriate cultivars. The 1979 survey showed that about 1% and 2% of farmers grew mixtures of winter wheat and spring barley cultivars respectively.

DEPLOYMENT OF CULTIVARS BETWEEN REGIONS AND COUNTRIES

Diversifying cultivars between regions or countries is also desirable. In North America, oat cultivars possessing different resistances to *Puccinia coronata* (crown rust) are deployed in three regions, between northern Mexico and the prairies of Canada (Frey *et al.*, 1977). Although no formal system for diversification on this scale yet exists in north-west Europe, there is the potential for such a system to limit the spread of some airborne pathogens including, for example, *Puccinia graminis* (black rust), spores of which can be dispersed from Spain to England (Ogilvie & Thorpe, 1958) a distance of about 1500 km.

CONCLUSIONS

Information already available to farmers undoubtedly helps them to make a rational choice of cultivars and to deploy those chosen to minimize the risk of yield loss due to disease. However, it is hoped that even better information can be provided in the future.

   Accurate plotting of isopaths will depend on obtaining data from more sites than at present. Some improvement might result if existing incomplete sets of data from other sites could be used. It would be valuable to extend the system to include stem-base pathogens, particularly *Pseudocercosporella herpotrichoides* (eyespot) of wheat and barley

   Replacement of the traditional 0-9 resistance ratings by "yield loss equivalents" should also be considered. These would indicate the maximum yield loss to be expected should a particular disease become prevalent. Thus a cultivar with a resistance rating of 8 might have a yield loss equivalent of 1% whereas one with a rating of 3 might have an equivalent of 10%. Farmers could then evaluate disease risk in terms of expected yield loss. This would be particularly useful when choosing between cultivars of approximately the same potential yield but different resistance.

   The cultivar diversification schemes are likely to be extended to incorporate other pathogens. The wheat scheme has more immediate potential as many of the cultivars also possess specific resistances effective against *Puccinia recondita* (brown rust) (Clifford *et al.*, 1979) and *E. graminis* (Bennett, 1979). Consideration is also being given to the most effective way of deploying resistance that is apparently race-non-specific, such as that commonly possessed by cereal cultivars against *R. secalis*, *Septoria nodorum* and *P. herpotrichoides* (Priestley, 1979).

   Cultivar mixtures are attractive because disease control is achieved with, apparently, very little cost to the farmer. They may also have the advantage of more stable yields than pure stands (Wolfe *et al.*, this volume). Further work is needed to determine if there are circumstances in which diversification within fields is more effective than between fields and *vice versa*. On theoretical grounds, diversification within fields seems likely to be more effective than diversification between fields in controlling those pathogens with short distance spore-dispersal mechanisms, such as those that are splash-dispersed.

   The potential for diversification on an international scale has hardly been considered in north-west Europe. At present much natural diversity still exists because until recently plant breeders tended to develop cultivars specifically for use in their own countries so that

many different types of resistance were exploited in relatively discrete areas. In recent years, there has been a growing tendency to
breed widely adapted cultivars for the international market. In the
short term this may be attractive economically but in the long term it
will tend to decrease diversity.

REFERENCES

Anon. (1979a) *Farmers' Leaflet No 8 - Recommended varieties of cereals
    1979*. National Institute of Agricultural Botany, Cambridge.
Anon. (1979b) *United Kingdom Cereal Pathogen Virulence Survey 1978
    Annual Report*, Cambridge. 66 pp.
Anon. (1979c) *The use of fungicides and insecticides on cereals 1979*.
    MAFF (Publications), Tolcarne Drive, Pinner, Middlesex HA5 2DT.
Anon. (1979d) *Recommended varieties of cereals 1979*. Publication No.
    45, The Scottish Agricultural Colleges.
Bennett F.G.A. (1979) Mildew of wheat. *United Kingdom Cereal Pathogen
    Virulence Survey 1978 Annual Report*, Cambridge, pp. 3-13.
Clifford B.C., Jones E.R.L. & Priestley R.H. (1979) Brown rust of wheat
    United Kingdom Cereal Pathogen Virulence Survey 1978 Annual Report,
    Cambridge, pp. 25-30.
Doodson J.K. (1976) Disease standards and diversification of varieties
    as a means of decreasing cereal diseases. *Proceedings of the 14th
    NIAB Crop Conference 1976*, pp. 32-9.
Frey K.J., Browning J.A. & Simons M.D. (1977) Management systems for
    host genes to control disease loss. *Annals of the New York Academy
    of Sciences* 287, 255-74.
King J.E. (1972) Surveys of foliage diseases of spring barley in
    England and Wales, 1967-70. *Plant Pathology* 21, 23-35.
King J.E. (1977a) Surveys of diseases of winter wheat in England and
    Wales, 1970-75. *Plant Pathology* 26, 8-20.
King J.E. (1977b) Surveys of foliage diseases of spring barley in
    England and Wales, 1972-75. *Plant Pathology* 26, 21-9.
Ogilvie L. & Thorpe I.G. (1961) New light on epidemics of black stem
    rust of wheat. *Science Progress* 49, 209-27.
Priestley R.H. (1978) The incidence of rust diseases in cereal cultivar
    trials in England and Wales 1957-1976. *Journal of the National
    Institute of Agricultural Botany* 14, 414-27.
Priestley R.H. (1979) The management of resistant varieties. *Pro-
    ceedings 1979 British Crop Protection Conference - Pests and Diseases*
    3, 753-60.
Priestley R.H. & Bayles R.A. (1979a) The incidence of powdery mildew in
    cereal cultivar trials in England and Wales 1957-1976. *Journal of
    the National Institute of Agricultural Botany* 15, 55-66.
Priestley R.H. & Bayles R.A. (1979b) The incidence of *Rhynchosporium
    secalis* and *Septoria* spp. in cereal cultivar trials in England and
    Wales 1957-1976. *Journal of the National Institute of Agricultural
    Botany* 15, 67-75.
Priestley R.H. & Byford P. (1979) Yellow rust of wheat. *United Kingdom
    Cereal Pathogen Virulence Survey 1978 Annual Report*, Cambridge,
    pp. 14-24.

Priestley R.H. & Wolfe M.S. (1977) Crop protection by cultivar diversification. *Proceedings 1977 British Crop Protection Conference - Pests and Diseases* 1, 135-40.

Rowe J. & Doodson J.K. (1976) The effects of mildew on yield in selected spring barley cultivars: a summary of comparative trials using fungicide treatments 1971-1975. *Journal of the National Institute of Agricultural Botany* 14, 19-28.

Wolfe M.S. (1978) Some practical implications of the use of cereal variety mixtures. *Plant Disease Epidemiology* (Ed. by P.R. Scott & A. Bainbridge), pp. 201-7. Blackwell Scientific Publications, Oxford.

Wolfe M.S. & Slater S.E. (1979) Mildew of barley. *United Kingdom Cereal Pathogen Virulence Survey 1978 Annual Report*, Cambridge, pp. 31-43.

Wolfe M.S. & Wright S.E. (1978) Mildew of barley. *United Kingdom Cereal Pathogen Virulence Survey 1977 Annual Report*, Cambridge, pp. 29-36.

The use of cultivar mixtures for disease control

M. S. WOLFE*, J. A. BARRETT† & J. E. E. JENKINS‡
*Plant Breeding Institute, Trumpington, Cambridge
†Genetics Department, Cambridge University, Cambridge
‡Agricultural Development and Advisory Service, Lawnswood, Leeds

INTRODUCTION

When powdery mildew resistant barley cultivars are grown on large areas they commonly become susceptible to specifically adapted genotypes of the pathogen (*Erysiphe graminis* f. sp. *hordei*). To delay or prevent this there is a need to present the pathogen with the greatest possible diversity of host resistance and this is now recognized in the United Kingdom in recommendations to diversify cultivars between fields (Anon., 1979; Priestley, this volume). However, to maximize the benefits of host diversity it is necessary to mix the host genotypes within a field so as to maximize cross inoculation between hosts. For within field diversity to be most effective, there should be the greatest possible differences in resistances between the component cultivars so that selection on the pathogen for the ability to reproduce well on one host component may lead to selection against that ability on others.

A decrease in disease in a mixture of cultivars does *not* depend on differences in race-specific resistance between the components. Disease may be decreased if the cultivars merely differ in resistance to an unspecialized pathogen population or a single genotype of a pathogen. In such a system, the amount of disease on a resistant cultivar will be relatively little affected by other, more susceptible cultivars. A susceptible cultivar surrounded by resistant plants will tend to have less disease than if it were grown alone (Jeger, 1979; Jeger *et al.*, this volume).

This observation is important because it implies that a mixture may be effective against unspecialized, non-target diseases, and that it may not select a "super-race" of the target disease. A race that is pathogenic on all components of a mixture is unlikely to be equally so on all of them; if there is a difference in host susceptibility, then the reaction of a mixture will tend towards that of the least affected component.

Cultivar mixtures can be easily exploited to provide a simple, cheap means of controlling important airborne diseases (Wolfe & Barrett, 1979,

1980) and will also insure against other diseases that occur sporadically. Their use in the European Economic Community is more feasible
now that the sale of mixtures of cereal cultivars is permitted within,
though not between, member states.

Multilines, in which the component cultivars are almost identical
except for particular resistances to the target pathogen, can be used
to control disease in a similar way (Browning & Frey, 1969) but the
uniformity of the genetic backgrounds of the host lines eliminates a
significant source of diversity of disease resistance. The multiline
approach is a conservative breeding strategy and may lead to difficulties in cultivar registration and multiplication. It may also be difficult to alter the composition of a multiline to take advantage of new
resistances.

DISEASE CONTROL IN MIXTURES OF SPRING BARLEY CULTIVARS

Control of disease by a cultivar mixture can be substantial and long-
lasting. Initially, when plants are subject only to infection from
external sources, there is little or no control but, as inoculum produced within a crop increases and becomes more important than that
produced elsewhere, the disease develops more slowly in a mixture than
in pure stands. The slowest rate of development, relative to that in
pure stands, is usually during the late seedling and early adult stages
when the crop canopy is still relatively open and a large proportion of
the spores is dispersed to host genotypes that they cannot infect.
Later, when the crop canopy has closed and plants may have several
tillers, spore movement between plants is limited so that disease development may be less affected by host heterogeneity.

In a three-cultivar mixture, powdery mildew infection is commonly
less than one half of the mean of the components grown alone (Wolfe,
1978; Wolfe & Barrett, 1979, 1980) (Table 1).

An experiment in 1979 showed that differences in unidentified
resistances of cultivars with the same identified resistance genes were
significantly and equally effective in reducing disease in the three
mixtures tested (Table 2). The unidentified resistances accounted for
about a quarter of the disease control attained with a mixture of three
cultivars differing in both identified and unidentified resistance
factors (Table 2). The effect of the unidentified resistances may have
been due to differences in resistance of the components to the same
pathogen race, or to each cultivar selecting its "own" sub-race from
the pathogen population.

Appropriate mixtures can also provide significant control of yellow
rust (*Puccinia striiformis*), brown rust (*Puccinia hordei*) and rhynchosporium (*Rhynchosporium secalis*).

Table 1. Powdery mildew (mean % area of 2nd and 3rd youngest leaves at GS* 70-80) on four barley cultivar mixtures grown at seven sites in 1979, compared with the means of their components grown in pure stands.

Site	Cultivars	Component mean (a)	Mixture (b)	$\underline{b}$ (%) a
		Mildew (%)		
1	Athos – Ark Royal – Lofa Abed	14.3	0.5	3
2	Athos – Georgie – Tyra	10.0	2.0	20
3	Athos – Georgie – Mazurka	9.0	1.4	16
4	Hassan – Midas – Wing	15.0	5.0	33
5	Hassan – Midas – Wing	14.3	5.0	35
6	Hassan – Midas – Wing	23.3	10.0	43
7	Hassan – Midas – Wing	8.0	3.0	38
Mean		13.4	3.8	28

* Zadoks *et al.* (1974)

Table 2. Powdery mildew (whole plot scores relative to the means of the components grown alone) on three barley cultivar mixtures, in each of which the components have the same identified resistance genes but different unidentified resistances, and a fourth mixture in which each cultivar has different identified and unidentified resistance genes.

Identified genes	Cultivars	Date 4/6	11/6	20/6	28/6
Mlg+Mlv	Abacus-Georgie-Sundance	100	83	82	47
Mlg+Mlas	Aramir-Athos-Porthos	116	83	74	34
Mla4/7	Ark Royal-Mazurka-Wing	102	72	68	64
(Mlg+Mlas)+ Mla4/7+(Mlg+ Mlv)	Athos-Mazurka-Sundance	33	36	31	22

THE YIELDS OF MIXTURES OF SPRING BARLEY CULTIVARS

In 1978 data were analysed from 26 different sites in England and
Scotland (Wolfe & Barrett, 1980). The trials differed in size and
design but allowed 47 comparisons between mixtures and their components
involving 37 different three-cultivar combinations from a total of 25
different cultivars representing 12 different groups of resistance
genes. The average yield of all mixtures was 106.5% of the weighted
mean of the components grown alone. At sites where mildew was considered important, the average yield of mixtures was about 109%; where the
disease was absent or considered unimportant, it was about 103% of the
mean of the components.

In 32 of the comparisons, all cultivars involved were on the National
Institute of Agricultural Botany Recommended List for 1978. The mean of
the highest yielding cultivars from each of these comparisons was 107.8%
and that of the mixtures was 106.5%, of the mean of all cultivars grown
alone. However, the highest yielding cultivars were not those that
would have been predicted from the Recommended List; their average yield
was only 102.3%.

Over all comparisons in which it occurred, no single cultivar yielded
as much as the mixtures in which it was included. Thus, mixtures of the
highest yielding cultivars often yielded more than the potential
expected from the Recommended List. The reason for increased yields of
the mixtures over the means of the components even at those sites where
mildew was absent or unimportant is not known; it may have been due to
control of other diseases, or to buffering of the mixtures against variations in the environment.

PATHOGEN POPULATION DYNAMICS IN PURE STANDS AND MIXTURES OF SPRING BARLEY CULTIVARS

The greater the amount of disease controlled by a cultivar mixture, the
more intense is selection of pathogen genotypes able to grow on all of
its components. The ability of the pathogen to respond to selection
appears to depend on the absolute rate of pathogen reproduction, spore
distribution, and the relative rate of reproduction of the different
pathogen populations on the different host components (Barrett, 1978,
1980). For example, if most spores reinfect the plants on which they
were produced, then pathogen genotypes best adapted to individual host
cultivars will be favoured: genotypes adapted to more than one cultivar
will decline in frequency if they have only a slightly inferior rate of
reproduction. At the other extreme, if all spores are redistributed at
random over all hosts, then selection may favour complex genotypes able
to grow on more than one host cultivar, even when these are less well
able to reproduce than is a simple genotype on the host to which it is
adapted.

In reality the distribution of spores probably lies somewhere
between these extremes, and is probably affected by plant growth and

habit, and weather. Because it is difficult to measure shifts in pathogen populations and the factors that cause them, particularly in cultivar mixtures, it is impossible to predict accurately the durability of disease control. However, computer simulations (Barrett, 1978, 1980) show that equilibria between simple and more complex races are unlikely to be established; if they do occur, they will probably be unstable or neutral.

Field results indicate that growing mixtures can lead to an increase in the frequency of pathogen races of intermediate complexity. Since the pathogen population is smaller, however, the number of such genotypes is less than that produced on an equivalent area of pure stands of the component cultivars (Wolfe & Barrett, 1979; Chin, 1979). Further, a complex pathogen race may not be highly pathogenic to all of the cultivars that it can attack but may be composed of genotypes with differential quantitative adaptation to each (Chin, 1979).

Some mixtures have given consistently good disease control but, so far, for only 5 years and in a small area; it is impossible to predict the response of the pathogen population if one mixture was to gain widespread popularity. Consequently, if mixtures do become widely used, there should be frequent changes in their range and composition. Limiting mixture composition to only three cultivars maximizes the potential for between-crop diversity and simplifies matching of available components for other characters.

THE FUTURE EXPLOITATION OF MIXTURES

Population dynamics of the pathogen

If cultivar mixtures become widely grown it will be important to monitor how well they control disease and how they affect the dynamics of the pathogen populations. If pathogen populations could be monitored over the whole of western Europe it might be possible to recommend mixtures with the best combinations of resistance factors against particular diseases on a regional basis.

Induced resistance

The mechanisms of powdery mildew control in cultivar mixtures include induced resistance (Chin, 1979). Although this is restricted to the immediate area of infection and, therefore, has little effect in one pathogen generation, it is cumulative and may account for 15% of the observed disease control later in the season. The magnitude of the induced resistance differs between cultivars (Chin, 1979) suggesting that appropriate choice of hosts to better exploit this phenomenon could increase the effectiveness of mixtures. Induced susceptibility also occurs but the size of this effect is much smaller.

Fungicide integration

Where fungicides are applied to seed to control foliar diseases, diversity in a mixture can be increased by, for example, applying the fungicide to a different host component in each year. Even if the fungicide cannot be wholly restricted to the treated cultivar the pathogen will be faced with a different problem each year. Integrating fungicide use with mixtures has further advantages in that cost will be reduced, since only one third of the seed needs to be treated, and there will be less risk of pollution and undesirable side-effects. Selection of fungicide-insensitive pathogen genotypes should also be decreased, which is especially important when so many new fungicides (triazoles, pyrimidines, imidazoles and morpholines) and at least one established compound (tridemorph), have a similar mode of action.

Other crops

Studies with spring barley suggest that cultivar mixtures should be effective in controlling airborne diseases of all intensively-grown, inbred crops. Indeed, preliminary experiments have shown similar benefits with both winter barley and winter wheat. Ideally, different resistance genes should be used in the winter and spring crops to ensure disruptive selection on the pathogen population between the two crops.

With winter wheat there may be practical difficulties in preparing large quantities of mixed seed in time for sowing. Also, the range of potentially useful mixtures of existing cultivars is limited since they have to provide control of powdery mildew, yellow rust and brown rust, whilst being compatible for date of maturity, herbicide insensitivity and milling quality. However, the potential development of genotypes with more durable resistance to *Puccinia striiformis* (Johnson, 1978, 1979, this volume) could provide a more reliable source of cultivars to increase the range of mixture components for controlling other diseases.

For feed grain, silage or chemical feedstock production, it may be worth further considering the potential of mixed cereal crops or dredge corn. In practice these mixtures may be extremely durable since complex races of foliar pathogens, able to attack more than one host species, are unlikely to arise.

Seed preparation

Commercial preparation of seed of winter cereals has to be completed in a short period so that the additional process of mixing would be difficult to introduce. The problem would be simplified if seed was multiplied in mixtures for a number of years depending on the quantities of grain required. However, shifts in composition of the mixtures affecting their performance might occur and these would be difficult to circumvent. A single year of multiplication as a mixture might provide an acceptable compromise.

Agronomy

There is little evidence from any crop that in the absence of disease
there are consistently large improvements in the yields of mixtures
compared with the means of their components (Trenbath, 1974). However,
this may be because all successful crop cultivars have been bred for
high yield and minimal response to environmental variation. Use of
cultivar mixtures does raise the possibility of selecting and combining
cultivars that are particularly well adapted to different, unpredictable
features of the environment. For example, a mixture of cultivars
adapted to either wet or dry conditions might perform better over a
number of sites and seasons than a single cultivar that performed
reasonably well under all moisture conditions.

It may also be possible to select mixture components that differently
exploit more constant features of the environment and so decrease inter-
plant competition. For example, cultivars, or species, that exploit
different regions of the soil or make use of light in different ways,
may exploit the environment more efficiently than closely-spaced plants
of almost identical structure and function.

Grain quality

For many characters the performance of a cultivar mixture would be
expected to equal the average of the components grown alone. The amount
of disease and, consequently, yield may depart greatly from this value
because there is an interaction of the host components with the pathogen. As grain quality is less affected by most diseases than is yield,
we can expect the quality of a mixture to depart little from the mean of
the components. Micromalting tests of barley grain have confirmed this
(Table 3).

Table 3. Nitrogen contents and malt extracts (from micromalting
tests) of grain from barley cultivar mixtures and the mean values
for their components grown alone, in 1975 and 1978.

Year		Grain nitrogen (%)	Soluble nitrogen (%)	Total extract (%)	Adjusted extract (1°/kg)
1975 (means of 6 mixtures)	components	2.13	0.432	65.6	267.2
	mixtures	2.09	0.447	65.9	266.0
1978 (means of 4 mixtures)	components	1.83	0.603	74.9	302.7
	mixtures	1.82	0.599	74.8	302.1

Because all of the different mixtures performed in this predictable way using a standard set of conditions, it seems likely that an optimum regime for malting a mixture could be easily determined from a knowledge of the requirements for the individual components.

In wheat mixtures, the grain quality of the mixtures may also be expected to be similar to the mean of the components grown alone. However, there may be a need to grow separate mixtures of hard and soft milling wheats.

ACKNOWLEDGEMENTS

Dr Barrett's work was supported by grants from the Agricultural Research Council to Prof. J.M. Thoday and Dr P. O'Donald.

REFERENCES

Anon. (1979) Annual Report of the UK Cereal Pathogen Virulence Survey 1978. UK Cereal Pathogen Virulence Survey Committee, Cambridge, 66 pp.

Barrett J.A. (1978) A model of epidemic development in variety mixtures *Plant Disease Epidemiology* (Ed. by P.R. Scott & A. Bainbridge), pp. 129-37. Blackwell, Oxford.

Barrett J.A. (1980) Pathogen evolution in multilines and variety mixtures. *Phytopathologische Zeitschrift* (in press).

Browning J.A. & Frey K.J. (1969) Multiline cultivars as a means of disease control. *Annual Review of Phytopathology* 7, 355-82.

Chin K.M. (1979) Aspects of epidemiology and genetics of the foliar pathogen, *Erysiphe graminis* f. sp. *hordei*, in relation to infection of homogeneous and heterogeneous populations of the barley host, *Hordeum vulgare*. Ph.D. thesis, University of Cambridge.

Jeger M. (1979) Studies of disease spread in heterogeneous cereal populations. Ph.D. thesis, University of Wales.

Johnson R. (1978) Practical breeding for durable resistance to rust diseases in self-pollinating cereals. *Euphytica* 27, 529-40.

Johnson R. (1979) The concept of durable resistance. *Phytopathology* 69, 198-9.

Trenbath B.R. (1974) Biomass productivity of mixtures. *Advances in Agronomy* 26, 177-210.

Wolfe M.S. (1978) Some practical implications of the use of cereal variety mixtures. *Plant Disease Epidemiology* (Ed. by P.R. Scott & A. Bainbridge), pp. 201-7. Blackwell, Oxford.

Wolfe M.S. & Barrett J.A. (1979) Disease in crops: controlling the evolution of plant pathogens. *Journal of the Royal Society of Arts* 127, 321-33.

Wolfe M.S. & Barrett J.A. (1980) Can we lead the pathogen astray? *Plant Disease* 64, 148-55.

Zadoks J.C., Chang T.T. & Konzak C.T. (1974) A decimal code for the growth stages of cereals. *Weed Research* 14, 415-421.

Effects of cereal cultivar mixtures on disease epidemics caused by splash-dispersed pathogens

M. J. JEGER*, ELLIS GRIFFITHS & D. GARETH JONES

Department of Agricultural Botany, University College of Wales, Aberystwyth, Dyfed

INTRODUCTION

The most widely grown temperate cereals (wheat, barley, oats) are inbred, and commercial cultivars are genetically uniform. Until recently, control of rusts and powdery mildew in such crops has been almost totally dependent upon the use of major-gene-resistant cultivars. However, to counter the rapid emergence of new pathotypes it has been necessary to introduce a succession of new cultivars containing different resistance genes. Consequently, scientists are considering alternative strategies, many of which involve more complex ways of deploying the available resistance genes. One such strategy is the deliberate introduction of genetic diversity into host-plant populations either by means of multilines (mixtures of near-isogenic lines differing in resistance genes) or by cultivar mixtures where the individual cultivars carry different resistance genes; both have been shown to decrease disease. Multilines have been used to control crown rust of oats (Browning & Frey, 1969; Frey et al., 1977; Browning & Frey, this volume) and cultivar mixtures to control powdery mildew of barley (Wolfe et al., 1976; Wolfe & Barrett, 1977; Wolfe, 1978; Wolfe et al., this volume).

Detailed studies of the role of genetic diversity in disease control and of the mechanisms involved have been restricted to host-pathogen systems in which the pathogen shows marked race specialization (Vanderplank, 1968; Leonard, 1969; Johnson & Allen, 1975; Barrett, 1978; Burdon, 1978; Chin, 1979). From these studies three processes which can contribute to the decrease in disease development have been identified. (1) *Inoculum dilution.* Spores produced on one cultivar or isogenic line are unable to infect most or all of the other components in the mixture. As a result much viable inoculum is ineffective and the progress of the epidemic is retarded. (2) *Induced resistance.* This results when attempted infection by spores of an avirulent patho-

---

* Present address: Plant Pathology Department, East Malling Research Station, Maidstone, Kent

Jenkyn J. F. & Plumb R. T. (1981) *Strategies for the Control of Cereal Disease*

type creates conditions which restrict development of a normally virulent pathotype. The converse effect (*induced susceptibility*) also occurs and the relative magnitude of these two effects is of obvious importance. With mildew on barley, induced resistance is quantitativel the most important and makes a significant contribution to disease control in cultivar mixtures (Chin, 1979). (3) *Plant density*. In a mixture, individuals susceptible to the same pathotype are separated an this may contribute to disease control (Burdon, 1978).

The successful operation of these particular mechanisms in decreasin disease depends as much on genetic diversity in the pathogen as in the host. If pathotypes capable of attacking all components in a mixture ("super races") arise then, it is argued, the advantages of host diversity are diminished. There is, therefore, an implication that hos diversity will be largely ineffective in decreasing diseases caused by unspecialized pathogens which have no race structure. However, some evidence (reviewed by Jeger, 1979) suggests that host diversity can als be effective in controlling unspecialized pathogens.

Resistance to unspecialized pathogens is often expressed as a slower rate of epidemic development. The rate of progress of an epidemic is dependent upon the interaction of three components of partial resistance namely infection frequency, latent period and sporulation (Parlevliet, this volume); their interaction is modified by environment. These components, in so far as they are host determined, are measurable and their resultant interaction can be examined in model systems. We have attempted to discover what happens when cereal cultivars, differing quantitatively in their components of partial resistance, are mixed together. We have adopted two approaches, (1) theoretical, by the construction of mathematical models and (2) experimental, using the host-pathogen combinations wheat/*Septoria nodorum* and barley/*Rhynchosporium secalis*.

MODELS OF EPIDEMICS IN MIXED STANDS

Two models have been developed to predict the development of an epidemic of an unspecialized pathogen in a mixture of cultivars. Both assume that a spore derived from one cultivar can infect either the same or a different component cultivar. The amounts of disease at a given time will depend upon a number of variables, of which infection frequency and sporulation are the most important because they interact in a mixture.

The first model considers disease progress as a sequence of "events" separated by latent periods. Each "event" consists of the sequence sporulation, dispersal and infection. A recurrence relationship can then be set up for both pure and mixed cultivar stands. In a simple 1:1 mixture of cultivars, amount of disease, as compared with that in the pure stands, can be predicted qualitatively. If the infection frequencies of the two cultivars A and B are denoted by F_a and F_b respectively and the sporulation rates by S_a and S_b, then if $F_a S_a = F_b S_b$ the amount of disease in the mixture will not differ from the arith-

metic mean of the amounts of disease in the pure stands. However, any
inequalities will modify the result. Thus if

$$F_a \cdot F_b S_b + F_b \cdot F_a S_a \; < \; F_a \cdot F_a S_a + F_b \cdot F_b S_b \; \ldots \ldots (i)$$

then disease in the mixed stand will be less than the arithmetic mean
of the pure stands. If the inequality is reversed then disease in the
mixed stand will exceed the arithmetic mean of the pure stands. A
simple modification of this model allows one to predict how disease
progress will be affected by different mixture ratios over any number of
"events".

The consequences of growing mixtures of cultivars with different
infection frequencies and sporulation rates are shown qualitatively in
Fig. 1. Benefits *(A)* can occur but, theoretically at least, certain
combinations of cultivars *(B)* would increase amounts of disease compared
to the pure-stand means *(C)*. However, for this to occur one cultivar
must have a lower infection frequency and permit higher sporulation than
the other. In practice, this is unlikely because we have no knowledge
of cultivars with opposition in infection frequency and sporulation of
sufficient magnitude.

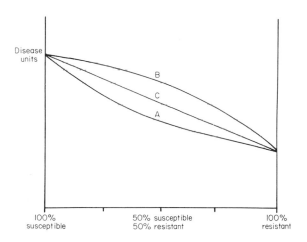

Figure 1. Consequences of growing mixtures of cultivars with different infection frequencies and sporulation rates. *A* results when
inequality (i) in the text is satisfied. *B* results when the inequality is reversed. *C* gives the arithmetic mean of pure stands.

It is possible to develop this model even further, taking into account
environmental effects, interval estimates of infection frequency and
sporulation rate, and changes in pathogen aggressiveness due to repeated
passage through the same or different host cultivars (Harrower, 1977).
However, the purpose of the model was to predict the likely consequences

of mixing before testing in the field and not to mimic an epidemic.
Hence, as an alternative to simulation using the discrete model, the
recurrence relationships established were written in the form of differential equations and investigated for analytical solutions.

Whether an exponential (no upper limit to disease) or logistic
(finite upper limit to disease) model is adopted there exist solutions
and, as with the first discrete model, these reduce to inequalities
involving the same two parameters. For the exponential model, again
using a 1:1 mixed stand, if

$$1 + \frac{F_b}{F_a}. \quad \sqrt{\exp{(F_b.S_b - F_a.S_a)}} < 1 + \frac{F_b}{F_a}. \quad \exp{(F_b - S_b - F_a.S_a)} \quad \ldots (ii)$$

then disease levels in the mixed stand will be less than the arithmetic
mean of the pure stands. Again, qualitative predictions can be made.
It is clear from the equations that lead to (ii) that disease in a mixed
stand will approximate to the geometric mean of the pure stands, and for
two non-negative numbers this is always less than the arithmetic mean.
For any amount of disease in pure stands it is possible to predict
amounts in mixed stands.

The models proposed here can also be modified and applied to systems
involving race-specialized pathogens and intergeneric mixtures, and
again predict benefits from mixing. Hence they do not conflict with the
models already proposed for those situations.

EXPERIMENTAL TESTING OF MODELS

The models described are simple and their use requires measurements of
only infection frequency and sporulation. Qualitative predictions of th
relative amounts of disease in mixed and pure stands depend on the
rankings of cultivars for these two characters. The usual prediction is
that disease will be decreased in mixed stands but predicting by how
much is more difficult and requires improvements in both the models and
the techniques for measuring infection frequency and sporulation. The
discrete model, in particular, is only valid when the component cultivars have similar latent periods.

Numerous estimates were made of infection frequency, latent period
and sporulation of S. nodorum on two spring wheat cultivars. The actual
values varied with both environment and plant age. The relative values
on the two cultivars also changed but the rankings of the cultivars
relative to each other did not. Initially, the mean values from several
experiments were used in the model and this predicted reasonably
accurately the results of an experiment testing mixtures of the two
spring wheat cultivars, Kolibri and Maris Butler. These two cultivars
give virtually identical infection frequencies and latent periods but
mean sporulation on cv. Kolibri was 1.8 times that on cv. M. Butler.

The qualitative prediction, using the discrete model, is that disease

in the mixture will be less than the mean of that in the pure stands.
The results of running a computer program for five "events" (infection
cycles) assuming sporulation rates for cv. Kolibri 1.5 and 2.0 times
those for cv. M. Butler are shown in Fig. 2. The observed amounts of
disease in replicated field plots each 4 m diameter, sown with different
proportions of the two cultivars, are also shown. The agreement between
predicted and observed values is reasonably good although the mixture
containing 25% of the more resistant cultivar (M. Butler) performed much
better than predicted.

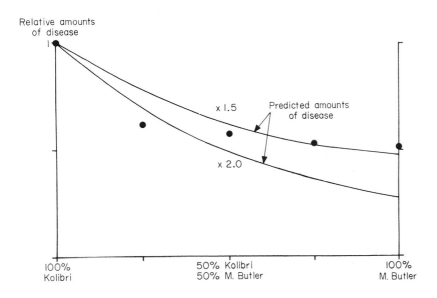

Figure 2. Relative amounts of disease observed (●) on cvs Kolibri
and Maris Butler and their mixtures compared with predicted amounts,
assuming sporulation on cv. Kolibri to be 1.5 or 2 times that on cv.
Maris Butler.

Experiments were also done with *S. nodorum* on winter wheat (cvs
Maris Huntsman and Maris Ranger) and *R. secalis* on winter barley (cvs
Hoppel and Maris Otter). If the inequalities in infection frequency
and sporulation are used as the basis for prediction, then pure stands
of cv. M. Ranger should have had very much more disease than pure stands
of cv. M. Huntsman but very little disease developed and differences
between the two cultivars were much smaller than predicted. The most
likely reason for this poor fit is inaccuracy in measurement of sporulation; in individual experiments values varied widely and a simple
arithmetic mean was probably inappropriate. Measuring sporulation of
R. secalis on barley was also difficult; on cv. Hoppel so few spores
were produced that reliable estimates could not be obtained. However,
a re-examination of the equations used in developing the analytical
model showed that specific values for sporulation were not required
provided estimates of infection frequency and of disease in pure stands

were available. This is because

$$\hat{y} = \{p + (1 - p)\lambda\} . (y_a)^P . (y_b/\lambda)^{1-p} \ \ldots (iii)$$

where, $\hat{y}$ = number of lesions in the cultivar mixture, p and $1-p$ are the proportions of cultivars A and B in the mixture, y_a and y_b are numbers of lesions on cultivars A and B in pure stands, and λ = ratio of infection frequencies F_b/F_a.

Predicted amounts of disease in mixtures containing different propor tions of the two cultivars, obtained from this equation, were compared with observed amounts (Table 1). The predicted values are closer to th

Table 1. Observed amounts of disease (% area affected on 3rd leaf) in field plots sown with mixtures or pure stands compared with amounts predicted using the model and "expected" amounts based on means of pure stands.

Winter wheat/ *S. nodorum*	Proportion Maris Huntsman/Maris Ranger				
	100/0	75 / 25	50 / 50	25 / 75	0/100
Observed	0.5	0.5 0.3 〰 0.8	0.3 0.7 〰 1.0	0.1 1.0 〰 1.1	3.6
Predicted		0.4 0.6 〰 1.0	0.3 1.4 〰 1.7	0.2 2.3 〰 2.5	
Expected		0.4 0.9 〰 1.3	0.3 1.8 〰 2.1	0.1 2.7 〰 2.8	

Winter barley/ *R. secalis*	Proportion Hoppel/Maris Otter				
	100/0	75 / 25	50 / 50	25 / 75	0/100
Observed	0.3	0.3 0.3 〰 0.6	0.4 1.1 〰 1.5	0.5 1.9 〰 2.4	13.7
Predicted		0.4 0.4 〰 0.8	0.6 1.5 〰 2.1	0.6 4.9 〰 5.5	
Expected		0.2 3.4 〰 3.6	0.2 6.9 〰 7.1	0.1 10.3 〰 10.4	

observed values than are the means of the pure stands but still over-estimate amounts of disease, especially where only 25% of the more resistant cultivar is present.

Yields were also taken from both diseased and disease-free plots. Although there were no significant differences there was a suggestion that yield was increased in mixed stands of wheat even in the absence of disease. By contrast, for barley the yield advantage in mixed stands was obtained only in the presence of disease. Clearly one cannot extra-polate from a decrease in disease to an increase in yield or assume that observed yield increases are necessarily due to disease control.

CONCLUSIONS

Heterogeneity in host populations can evidently retard epidemics of unspecialized pathogens such as *S. nodorum* and *R. secalis*. However, high yielding cultivars of wheat and barley with good resistance to these pathogens are available and at present there seems to be no justi-fication for introducing cultivar mixtures specifically aimed at their control. Cultivar mixtures may, on the other hand, be used to control stripe rust (*Puccinia striiformis*) of wheat and powdery mildew (*Erysiphe graminis*) of barley (Wolfe *et al.*, this volume). If the use of such mixtures became common agricultural practice then the performance of mixed stands against non-target diseases would be of importance and methods of predicting their performance would be desirable.

The models proposed here are, in essence, simple to use and, given correct rankings of cultivars for infection frequency and sporulation, it seems possible to predict qualitatively the effect of mixing on disease severity. Quantitative prediction is more difficult. Even if the models are themselves correct and robust, the problem of assigning values for infection frequency and sporulation on individual cultivars remains. Improvements in prediction will certainly require improve-ments in methods of measuring the components of partial resistance.

It is also evident that the relationships between disease control and yield benefits in mixtures are far from simple. Intervarietal competition, even in the absence of disease, makes prediction of mixture yields difficult if not impossible. In these circumstances empirical evaluation may well be the best that can be achieved in practice.

REFERENCES

Barrett J.A. (1978) A model of epidemic development in variety mixtures. *Plant Disease Epidemiology* (Ed. by P.R. Scott & A. Bainbridge), pp. 129-37. Blackwell Scientific Publications, Oxford.

Browning J.A. & Frey K.J. (1969) Multiline cultivars as a means of disease control. *Annual Review of Phytopathology* 7, 355-82.

Burdon J.J. (1978) Mechanisms of disease control in heterogeneous plant populations - an ecologist's view. *Plant Disease Epidemiology*

(Ed. by P.R. Scott & A. Bainbridge), pp. 193-200. Blackwell
Scientific Publications, Oxford.

Chin K.M. (1979) Aspects of the epidemiology and genetics of the foliar
pathogen, *Erysiphe graminis* f. sp. *hordei* in relation to infection of
homogeneous and heterogeneous populations of the barley host
(*Hordeum vulgare*). Ph.D. thesis, University of Cambridge.

Frey K.J., Browning J.A. & Simons M.D. (1977) Management systems for
host genes to control disease loss. *Annals of the New York Academy
of Sciences* 287, 255-74.

Harrower K.M. (1977) Virulence changes in an isolate of *Leptosphaeria
nodorum* to wheat cultivars. *Australian Plant Pathology Society
Newsletter* 6, 42-4.

Jeger M.J. (1979) Studies on disease spread in heterogeneous cereal
populations. Ph.D. thesis, University of Wales.

Johnson R. & Allen D.J. (1975) Induced resistance to rust disease and
its possible role in the resistance of multiline varieties. *Annals
of Applied Biology* 80, 359-63.

Leonard K.J. (1969) Factors affecting rates of stem rust increase in
mixed plantings of susceptible and resistant oat varieties.
Phytopathology 59, 1845-50.

Vanderplank J.E. (1968) *Disease resistance in plants*. Academic Press,
London. 206 pp.

Wolfe M.S. (1978) Some practical implications of the use of cereal
variety mixtures. *Plant Disease Epidemiology* (Ed. by P.R. Scott &
A. Bainbridge) pp. 201-7. Blackwell Scientific Publications, Oxford.

Wolfe M.S. & Barrett J.A. (1977) Population genetics of powdery mildew
epidemics. *Annals of the New York Academy of Sciences* 287, 151-63.

Wolfe M.S., Wright S.E. & Minchin P.N. (1976) Effects of variety
mixtures. *Report of the Plant Breeding Institute for 1975*, pp. 130-

Section 2
Chemical Control

Chairman's comments

T. F. PREECE
Agricultural Sciences Building,
The University, Leeds

Twenty-five years ago the application of chemicals to cereal crops meant applying fertilizers and herbicides. The only chemicals used for disease control were organo-mercury compounds applied to the seed before planting. Today, fungicides are widely used for leaf disease control in cereal crops. Why is this, and how is it that the aspects of chemical control of cereal diseases, which occupy one-third of this volume, have arisen at all in the technology of cereal growing?

The answer to this question is that 25 years ago no one knew what the losses caused by leaf-infecting fungi of cereals were - even approximately. Inspired by W.C. Moore, and co-ordinated by E.C. Large, field work done in England and Wales by advisory plant pathologists first quantified grain losses caused by *Erysiphe graminis* and, sub-sequently, other cereal diseases.

In financial terms the losses due to leaf diseases of cereals alone in England and Wales are equivalent in value to about £100 000 000 per annum. We have to reduce such losses by all the means at our disposal, and this includes use of the best fungicides we have available.

The following chapters consider some aspects of the present use of chemicals to control cereal diseases, representing the efforts of many plant pathologists in research, in advisory work and in the pesticide industry.

The deployment of fungicides in cereals

R. J. COOK*, J. E. E. JENKINS† & J. E. KING‡

*Agricultural Development and Advisory Service, Ty Glas Road, Cardiff
†Agricultural Development and Advisory Service, Lawnswood, Leeds
‡Ministry of Agriculture, Fisheries and Food,
Harpenden Laboratory, Hertfordshire

INTRODUCTION

Fungicides are being increasingly used on cereals in the United Kingdom (UK). They are used either routinely, on a calendar or growth stage basis (prophylactic treatment), or in response to an indication of an imminent risk of significant disease development. Some cereal diseases are adequately controlled by husbandry or the use of resistant cultivars but even where these are not effective, epidemics will only occur if other factors, such as the weather, are favourable. Thus, not all cereal crops will benefit from treatment with fungicides, so there is a need to develop methods which will assist farmers to make decisions on their cost-effective use. Because cereals occupy such a large area (3.3 M ha in England and Wales, which is 75% of the arable area and 35% of the total farm area excluding rough grazings), use of pesticides should also take into account their effects on the environment, which have political and social as well as biological implications.

DISEASE INCIDENCE AND SEVERITY

An essential first stage in developing a strategy for the deployment of fungicides is a national assessment of the incidence and severity of diseases and the grain losses they cause.

Surveys of spring barley in England and Wales, have shown that at the milky ripe growth stage (GS 75, Zadoks et al., 1974) mildew (Erysiphe graminis DC) is usually the most frequent and severe leaf disease, followed by brown rust (Puccinia hordei Otth.) or leaf blotch (Rhynchosporium secalis (Oudem.) J.J. Davis) and yellow rust (P. striiformis West.) (King, 1977a). Similar surveys of winter wheat (King, 1977b, 1977c) have shown that mildew is the most severe leaf disease in dry seasons, but Septoria nodorum Berk. (Leptosphaeria nodorum Müller) is more important in wet seasons. Relatively severe yellow rust and brown rust (P. recondita Rob. & Desm.) of wheat have only been associated with the widespread cultivation of particularly susceptible cultivars. Eyespot (Pseudocercosporella herpotrichoides

Jenkyn J. F. & Plumb R. T. (1981) *Strategies for the Control of Cereal Disease*

(Fron) Deighton) has been recorded since 1975 and symptoms sufficiently
severe to reduce yield of wheat have been observed on up to 2% of stems
The proportion of crops affected by barley yellow dwarf virus (BYDV)
varied considerably from year to year; ranging from 10% (1979) to 89%
(1967) in barley and, from 7% (1979) to 75% (1971 & 1974) in wheat.

LOSSES

The survey data have been used to estimate percentage losses in yield
using formulae relating loss to severity of barley and wheat mildew
(Large & Doling, 1962, 1963), barley leaf blotch (James et al., 1968),
and brown rust (King & Polley, 1976), wheat septoria (King, 1977b),
yellow rust (King, 1976) and eyespot (Scott & Hollins, 1978).

    Using Ministry of Agriculture data for England and Wales, the total
values of the 1978 crops (using 1979 prices) were estimated at £548M
for winter wheat and £517M for spring barley. Based on these values,
the disease of greatest economic importance over the period of the
surveys (1967-79) has been barley mildew with losses averaging £43M
and ranging from £19M in 1977 to £72M in 1968, although the latter
estimate is for a year before fungicides were available to control
barley mildew. Wheat mildew was the second most damaging disease with
estimated losses ranging from £8M in 1977 to £24M in 1973 and averaging
£14.5M. Other diseases have been more sporadic, the only ones causing
losses approaching that of barley mildew being septoria of wheat in
1972 (£41M) and brown rust of barley in 1970 (£40M).

DISEASE RISK

The losses caused by diseases demonstrate the potential value of fungicides on cereals but it will almost invariably be unnecessary and
wasteful to spray all crops. Only crops with sufficiently severe
disease to make treatment economically worthwhile should be sprayed.
The economics of spraying depend on the chemical used, method of
application, amount of damage caused in application (e.g. wheeling
damage causes losses of up to 4% in crops treated after flag leaf
emergence), machinery depreciation, fuel and labour charges and grain
value. Using survey data from untreated crops, the proportion of crops
likely to have justified treatment at different costs may be estimated
(King, 1977c) (Table 1).

    The proportion of barley crops justifying treatment for barley
mildew is high and it is estimated that the average crop is likely to
repay treatment costs as high as £40/ha in 26 cases out of 100. For
other diseases the survey data suggest that few crops justify treatment.
Few treatments cost as much as £40/ha; most cost £10-20/ha and data
indicate that the proportions of crops treated at these costs more than
match the proportions of crops at risk from diseases for which fungicides are available. For example in 1979, when broad spectrum
fungicides were widely used, 49% of barley and 52% of wheat crops were

treated with products capable of controlling the diseases shown in
Table 1.

Table 1. Percentages of untreated crops sampled in the winter
wheat and spring barley surveys where disease severities indicated
losses sufficient to repay control by fungicidal sprays at 1979
costs.

Disease	Number of crops examined	Percentage repaying treatment cost (£/ha) of			
		>10	>20	>30	>40
Barley mildew	2305	64	45	34	26
leaf blotch	2305	7	3	2	1
brown rust	2305	15	10	7	5
Wheat mildew	2629	18	9	6	4
septoria	2629	13	8	6	4
yellow rust	2629	3	2	1	1
eyespot	1214	8	4	3	2

CRITERIA FOR FUNGICIDE USE

The survey data provide information on national losses, the extent of
fungicide usage and the proportions of crops which are economically
worth treating with fungicide in England and Wales. It does not
necessarily follow, however, that fungicides were only applied to those
crops which were economically worth treating. To avoid unnecessary use
of pesticides the farmer or adviser must be able to identify crops at
risk. Several attempts have been made to do this. In the UK, Polley &
Smith (1973) developed criteria for timing the application of sprays to
control spring barley mildew based on the identification of weather
conditions which favour spore production. In practice these criteria
have not been reliable because other factors (e.g. inoculum level and
host resistance) are equally important in determining the development
of the epidemic. Similarly other forecasting schemes based on meteorological data (Polley & Clarkson, 1978) may help to identify periods
when crops need protection, but they cannot be used alone to determine
fungicide use.

However, such weather criteria have been more useful in the case of
septoria diseases of wheat where associations between disease and
weather patterns appear to form a useful basis for spray prediction
schemes. For example wet weather at or near heading is commonly
associated with severe development of septoria (Obst, 1977; Shaner &
Finney, 1976). Tyldesley & Thompson (1980), using wheat survey data,

have shown a relationship between the severity of disease caused by
S. nodorum at the milky ripe growth stage (GS 75) and the number of
days with more than 1 mm of rain in the second half of May and the
first half of June. In the UK the best yield responses have been
obtained when fungicides were applied immediately following weather
periods favourable to disease development during or just before the
period from flag leaf to ear emergence (GS 39-58), even though at the
time of application there may be little disease (Cook, 1977).

Data from epidemiological studies have been used to develop methods
for identifying potential epidemics and these have sometimes been
linked with fungicide use. Burleigh *et al.* (1972), working with wheat
brown rust in the USA, used step-wise multiple linear regressions and
identified six biological and meteorological variables which explained
most of the variation in disease progress. They were able to predict
rust severity within 1, 3 and 12% of that observed respectively 14, 21
and 30 days later. The most useful variable for prediction was disease
severity, though both biological and meteorological data were necessary
to provide acceptably accurate predictions. The object of this work
was to predict epidemics of brown rust so that fungicides could be
used in time to reduce crop loss but this has not been tested because
fungicides which are economic and effective are not cleared for use on
cereals in the USA.

Zadoks (1971) suggested the use of simulation modelling to predict
epidemics of wheat yellow rust and recently Teng *et al.* (1978) have
developed this method to predict epidemics of barley brown rust in New
Zealand and consequent crop losses. They have proposed that the model
could be applied to individual crops and envisage a direct link between
the farmer and the computer. Weather, crop growth and disease would be
monitored and the computer used to assess the cost/benefit of fungicide
use at any growth stage of the crop. A similar model developed for
S. nodorum by Rapilly & Jolivet (1976) recognizes the important need
to take account of differences in cultivar susceptibility but has not
yet been used to predict when fungicides are needed.

These complex models require large amounts of detailed information
about the weather and the crop. A method to predict eyespot infection
developed in the Federal Republic of Germany by Fehrmann & Schrödter
(1973) uses temperature and relative humidity. This has been used by
the German Weather Service since 1974 to aid timing of sprays of
carbendazim fungicides, although recently Fehrmann & Weihofen (1978)
suggested that sprays should be timed according to crop growth stage.
This follows the practice developed in the UK (Anon., 1979), Switzerland
(Vez & Gindrat, 1976), Schleswig-Holstein (Bockmann & Effland, 1975)
and France (Maumene *et al.*, 1979) where the probability that a crop
may be damaged by eyespot is estimated at GS 30-31 by using such
factors as plant density, sowing date, disease incidence, rotation,
cultivar susceptibility and soil type.

Recommendations developed by plant pathologists of the Agricultural
Development and Advisory Service (ADAS) to aid timing of fungicides to

control the main cereal diseases in England and Wales (Anon., 1979) recognize the influence of weather and husbandry but are based mainly on critical disease severity in the crop.

For spring barley mildew, data from over 50 experiments showed that the highest yield response to a single spray occurred if the spray was applied as soon as 3% of the area of the oldest green leaves was affected by mildew (Jenkins & Storey, 1975). Farmers are therefore advised to spray as soon as this level of disease occurs. With other pathogens the threshold level is adjusted in accordance with cultivar susceptibility. Thus with yellow rust of winter wheat (Anon., 1979), highly susceptible cultivars should be sprayed as soon as the disease is noticed but with moderately susceptible cultivars treatment may be delayed until 1% of the area of the top leaf is affected.

EFFECTS OF FUNGICIDES ON YIELD

Large yield increases can be expected when fungicides provide good disease control but there are examples of fungicides giving yield increases when disease is apparently slight or when significant control is not obtained. Jenkins et al. (1972), for example, recorded such an effect with frequent applications of the broad spectrum fungicides zineb and lime sulphur to spring barley. Late season applications of carbendazim/dithiocarbamate mixtures in 118 experiments on winter wheat done by ADAS plant pathologists between 1970 and 1978 (Cook, 1980) gave an average yield increase of 0.185 t/ha (3.3%) even though disease was

Table 2. Distribution (%) of responses in winter wheat experiments sprayed with carbendazim plus maneb or mancozeb. (Data for England and Wales, E&W, based on ADAS experiments; data for France,F, after Poussard et al., 1977).

Year	Number of values		Yield responses (t/ha) greater than					
	E&W*	F†	0.1		0.2		0.3	
			E&W	F	E&W	F	E&W	F
1973	54	92	76	76	61	55	43	41
1974	45	117	46	69	37	48	28	38
1975	55	138	56	60	27	43	16	22
1976	33	73	39	38	21	14	12	4
1977	59	–	56	–	37	–	17	–
1978	46	–	63	–	54	–	52	–
Mean	48	105	57	61	40	40	28	26

* Sprayed at GS 39–75; most between GS 39 and 59.
† Sprayed at GS 59.

slight or absent. Similar results have been reported from West Germany
(Fehrmann et al., 1978) and France (Poussard et al., 1977) where the
frequency of economic yield response to treatment was similar to that
obtained in the ADAS experiments (Table 2). In both sets of data
about 60% of late season treatments gave responses of greater than 0.1
t/ha; the equivalent of a treatment cost of about £10/ha. The causes o
yield increases obtained in the absence of significant amounts of
recognized diseases and the conditions under which they occur are not
known. There could be direct effects of the fungicides on host physiology such as the cytokinin effects of carbendazim fungicides
(Mukhopadhyay & Bandopadhyay, 1977), or activity against fungi such as
Alternaria spp., Cladosporium spp. and Botrytis cinerea Pers. ex Pers.
not normally recognized as pathogens (Dickinson, this volume).

Although these observations are not necessarily representative of
commercial crops, even small yield increases obtained when fungicides
are applied in the absence of significant amounts of disease need to
be considered when making a decision whether or not to spray.

CONCLUSIONS

The survey data for England and Wales suggest that the frequency of an
economic response is high for fungicides used to control spring barley
mildew but usually low for those used against other diseases. Thus
routine treatment against spring barley mildew can be justified at
present costs but it is sensible to apply fungicides against other
diseases only when economic damage threatens. However, many farmers
prefer to use fungicides routinely on a fixed schedule, based on
growth stages or time intervals. This simplifies management and
avoids time-consuming procedures such as frequent crop inspection. It
also avoids the need to take decisions which often require immediate
action and complicate management by requiring that a large area be
sprayed in a short time, when weather or soil conditions might be
unfavourable or other tasks on the farm have a high priority. Where
diseases are consistently expected to cause significant yield losses
routine treatment may be cost effective. This approach might also be
justified where high input, high yielding crops are grown and farmers
consider that some protection against disease is necessary. However,
where routine treatments cannot be justified the farmer needs simple
criteria to assist him in making decisions on fungicide use.

Equations of the kind developed by Burleigh et al. (1972) may be
useful in predicting disease over large areas, especially where environmental conditions and cultivars are relatively uniform. However,
they are impracticable for independant use by farmers. Similarly the
method proposed by Teng et al. (1978), which needs a large input of
data from the farmer and access to a computer, is likely to be
difficult to apply, at least in the immediate future. This is recognized in New Zealand where the immediate objective (R. Close, personal
communication) is to encourage farmers to use criteria such as disease
severity in the crop and sowing date. The assessment of disease

severity will probably be based on the incidence of affected leaves
(James & Shih, 1973) which is easier for a farmer to use than an
assessment of percentage leaf area affected.

In England and Wales the aim of ADAS plant pathologists has been to
develop simple criteria that the farmer can use in the field (Anon.,
1979). They have been derived empirically and involve the inspection
of individual crops. This is essential because disease development in
the UK is so variable. The decision to spray is determined by a
threshold of disease severity and cultivar susceptibility. These
factors alone are adequate for most diseases but for septoria in
winter wheat and possibly rhynchosporium in barley there is also a
need to take into account periods of wet weather at specific growth
stages.

Whatever the disease, the decision to spray or not is made, consciously or subconsciously by assessing the risks and the likely outcome
of different actions. Formalised decision making processes have been
applied to the use of fungicides for control of septoria on wheat by
Webster & Cook (1979). They used judgemental probabilities to estimate
the yield response to fungicide application for different sets of
simple, readily-identified field conditions. This method integrates
results of spray trials and the experience of advisers to provide a
simple guide before epidemiologically derived criteria are available.
Simple decision trees can be easily made and distributed to farmers
or their advisers. These use the specialist's experience but he does
not need to be present for decisions to be made. Furthermore they
identify occasions when sprays are unlikely to be needed.

Decisions on the use of fungicides should be based on individual
field conditions. However, disease information on a local and a
national basis can alert farmers and their advisers of the need to
inspect crops, particularly of known susceptible cultivars. With the
aid of computer-based television or telephone systems this information
can be quickly brought up to date or modified. Computers using
predictive models and linked to specific crops are not yet practicable
but the development of programs to warn of disease risk periods is
possible, especially as computer usage on the farm becomes more
widespread. There is clearly a need for more and better information
on which to base spray decisions predicting both the likely response
to a fungicide and the best time to apply it.

REFERENCES

Anon. (1979) The use of fungicides and insecticides on cereals 1979
    (CDP21). *Ministry of Agriculture, Fisheries and Food (Publications)*,
    Pinner. 73 pp.
Bockmann H. & Effland H. (1975) Über die Grundlagen einer prognose
    der Fusskrankheiten des Weizen. *Aktuelles aus Acker und Pflanzenbau*,
    6, 101-14.
Burleigh J.R., Eversmeyer M.G. & Roelfs A.P. (1972) Development of

98   R.J. COOK, J.E.E. JENKINS & J.E. KING

linear equations for predicting wheat leaf rust. *Phytopathology* 62, 947-53.

Cook R.J. (1977) Effect of timed fungicide sprays on yield of winter wheat in relation to *Septoria* infection periods. *Plant Pathology* 26, 30-4.

Cook R.J. (1980) Effects of late season fungicide sprays on yield of winter wheat. *Plant Pathology* 29, 21-7.

Fehrmann H. & Schrödter H. (1973) Control of *Cercosporella herpotrichoides* in winter wheat in Germany. *Proceedings 7th British Insecticide and Fungicide Conference* 1, 119-26.

Fehrmann H., Reinecke R. & Weihofen V. (1978) Yield increase in winter wheat by unknown effects of MBC fungicides and captafol. *Phytopathologische Zeitschrift* 93, 359-62.

Fehrmann H. & Weihofen V. (1978) Aktuelle Forschungsprobleme zur chemischen Bekämpfung von *Cercosporella herpotrichoides* in Weizen. *Zeitschrift für Pflanzenkrankheiten und Pflanzenschutz* 85, 142-9.

James W.C., Jenkins J.E.E. & Jemmett J.L. (1968) The relationship between leaf blotch caused by *Rhynchosporium secalis* and losses in grain yield of spring barley. *Annals of Applied Biology* 62, 273-88.

James W.C. & Shih C.S. (1973) Relationship between incidence and severity of powdery mildew and leaf rust on winter wheat. *Phytopathology* 63, 183-7.

Jenkins J.E.E., Melville S.C. & Jemmett J.L. (1972) The effect of fungicides on leaf diseases and yield in spring barley in South-West England. *Plant Pathology* 21, 49-58.

Jenkins J.E.E. & Storey I.F. (1975) Influence of spray timing for the control of powdery mildew on the yield of spring barley. *Plant Pathology* 24, 125-35.

King J.E. (1976) Relationship between yield loss and severity of yellow rust recorded on a large number of single stems of winter wheat. *Plant Pathology* 25, 172-7.

King J.E. (1977a) Surveys of foliar diseases of spring barley in England and Wales, 1972-75. *Plant Pathology* 26, 21-9.

King J.E. (1977b) Surveys of diseases of winter wheat in England and Wales, 1972-75. *Plant Pathology* 26, 8-20.

King J.E. (1977c) The incidence and economic significance of diseases in cereals in England and Wales. *Proceedings 1977 British Crop Protection Conference - Pests and Diseases* 3, 677-87.

King J.E. & Polley R.W. (1976) Observations on the epidemiology and effects on grain yield of brown rust on spring barley. *Plant Pathology* 25, 63-73.

Large E.C. & Doling D.A. (1962) The measurement of cereal mildew and its effect on yield. *Plant Pathology* 11, 47-57.

Large E.C. & Doling D.A. (1963) Effect of mildew on yield of winter wheat. *Plant Pathology* 12, 128-30.

Maumene J., Poussard C. & Prevot J.P. (1979) Piétin-verse: Méthode pratique de prévision des risques de dégâts. *Perspectives Agricoles* 26, 60-3.

Mukhopadhyay A.N. & Bandopadhyay R. (1977) Cytokinin-like activity of carbendazim. *Pesticides* 11, 24-5 & 28.

Obst A. (1977) Untersuchungen zur Epidemiologie, Schadwirkung und Prognose der Stelzenbräune (*Septoria nodorum*) des Weizens. *Bayer-*

isches Landwirtschaftliches Jahrbuch <u>54</u>, 72–117.

Polley R.W. & Smith L.P. (1973) Barley mildew forecasting. *Proceedings 7th British Insecticide and Fungicide Conference* <u>21</u>, 373–8.

Polley R.W. & Clarkson J.D.S. (1978) Forecasting cereal disease epidemics. *Plant Disease Epidemiology.* (Ed. by P.R. Scott & A. Bainbridge), pp. 141–50. Blackwell, Oxford.

Poussard C., Maumene J., Govet J.P. & Lescar L. (1977) Les traitments fongicides du blé tendre d'hiver en course de végétation. *Lutte contre les maladies et les ravageurs des céréales. Journée d'Étude, 26 Jan. 1977,* 169–77. *Institute Techniques des Céréales et des Fourrages,* Paris.

Rapilly F., & Jolivet E. (1976) Construction d'un modèle (Episept) permettant la simulation d'une épidémie de *Septoria nodorum. Revue statistique Appliquée* <u>24</u>, 31–59.

Scott P.R. & Hollins T.W. (1978) Prediction of yield loss due to eyespot in winter wheat. *Plant Pathology* <u>27</u>, 125–31.

Shaner G. & Finney R.E. (1976) Weather and epidemics of Septoria leaf blotch of wheat. *Phytopathology* <u>66</u>, 781–5.

Teng P.S., Blackie M.J. & Close R.C. (1978) Simulation modelling of plant diseases to rationalize fungicide use. *Outlook on Agriculture* <u>9</u>, 273–7.

Tyldesley J. & Thompson N. (1980). Forecasting *Septoria nodorum* on winter wheat in England and Wales. *Plant Pathology* <u>29</u>, 9–20.

Vez A. & Gindrat D. (1976) Opportunité de la lutte chimique contre le piètin-verse du blé. *Révue Suisse d'Agriculture* <u>8</u>, 33–8.

Webster J.P.G. & Cook R.J. (1979) Judgemental probabilities for the assessment of yield response to fungicide application against *Septoria* of winter wheat. *Annals of Applied Biology* <u>92</u>, 39–48.

Zadoks J.C. (1971) Systems analysis and the dynamics of epidemics. *Phytopathology* <u>61</u>, 600–10.

Zadoks J.C., Chang T.T. & Konzak C.F. (1974) A decimal code for the growth stages of cereals. *Weed Research* <u>14</u>, 415–21.

Timing fungicides for eyespot control

V. W. L. JORDAN & HILARY S. TARR

Long Ashton Research Station, Bristol

INTRODUCTION

Eyespot (*Pseudocercosporella herpotrichoides* (Fron) Deighton) can decrease yield either directly, or indirectly if it causes lodging or straggling (Scott & Hollins, 1974). Losses in winter wheat are mainly attributable to severe lesions (Ponchet, 1959; Scott & Hollins, 1974). By contrast, pot experiments with winter barley have shown that even slight infection can cause significant decreases in grain number per ear and in 1000 grain weight, but in the absence of lodging the more severe the lesion the greater the loss (Jordan & Tarr, 1979).

The carbendazim-generating fungicides are currently the most effective against eyespot. In Europe, the optimum time to apply single sprays in the spring is between the end of tillering and the appearance of the first node (Fehrmann & Schrödter, 1973; Hampel & Löcher, 1973; Rule, 1975; Taylor & Waterhouse, 1975). However, although fungicides applied at the optimum time often considerably decrease the percentage of diseased plants, yields are frequently only slightly increased. In North America Powelson & Cook (1969) demonstrated almost complete control of eyespot by fungicides applied either in November, March or April, whereas in other trials spring applications were found to be the most effective (Bruehl & Cunfer, 1972; Huber & Mulanax, 1972).

It seems to be more important to correctly time sprays, which may only decrease disease severity, than to eliminate infection (Fehrmann & Schrödter, 1973). However, prophylactic use of fungicides is often uneconomic and seldom justified. A better understanding of the biology of the pathogen may help to predict when sprays are required and improve their timing.

INOCULUM SOURCES

Primary eyespot infections develop from spores produced on infected culm bases remaining on or in the soil from previous crops but little is known about the factors affecting spore dispersal or about the amount of infective debris which constitutes a disease risk.

In 1976, on land that had no previous history of cereal crops, the effect of inoculum density on eyespot severity was tested by placing 1, 5, 10, 25, 50 or 100 eyespot-infected straws at random in 9 m^2 plots of winter barley cv. Hoppel or winter wheat cv. Maris Huntsman on 5 January (Growth Stage (GS) 24: Zadoks et al., 1974). On winter wheat, eyespot symptoms in plots with infected straws were significantly more severe at harvest than in uninoculated plots. On winter barley, by contrast, disease severity was significantly increased only in plots inoculated with 25 or more infected straws. Nevertheless, there was often more disease where only one infected straw was introduced than in uninoculated plots. Obviously very little infected stubble can, in suitable weather, cause a damaging attack. Burning stubble may be ineffective in preventing survival of eyespot (Slope et al., 1970) because the small portions that remain often contain eyespot lesions able to sporulate.

The importance of secondary spread in the epidemiology of eyespot is not clear. Glynne (1953) reported that eyespot lesions on young wheat plants in the field frequently produced conidia in the spring, but that conidia were not produced on mature plants. Ponchet (1959) attributed most spring infections to secondary inoculum, but considered that, in France, damage from these later infections was not serious. We have failed to obtain spores from lesions on the growing crop in south-west England.

SPORE DISPERSAL

Dispersal of eyespot spores from an inoculum source (15 infected straws) in the centre of a circular plot of winter wheat was monitored using modified rotorod samplers (O.J. Stedman, personal communication) 10 cm above soil level. Wind during periods of rain, when sporulation was most active, was predominantly from the west and conidia of P. herpo-trichoides were caught up to 198 cm from the source. Some infection occurred up to 180 cm to the east of the source but most was at 74 cm to the east (Fig. 1).

SPORE PRODUCTION

In experiments between autumn 1975 and summer 1979, the period over which infected debris produced spores was measured by collecting rainwater which had washed over eyespot-infected wheat culm bases (cv. Maris Huntsman) held above a funnel near ground level. In three of the four seasons (1975/6, 1976/7 and 1978/9) sporulation had ceased by early April at the latest (Fig. 2). By contrast, in 1977/8, numbers declined markedly during late March but then increased during April and May. Monitoring inoculum release might thus enable us to predict years when spring cereals exposed to infected debris are at risk from eyespot.

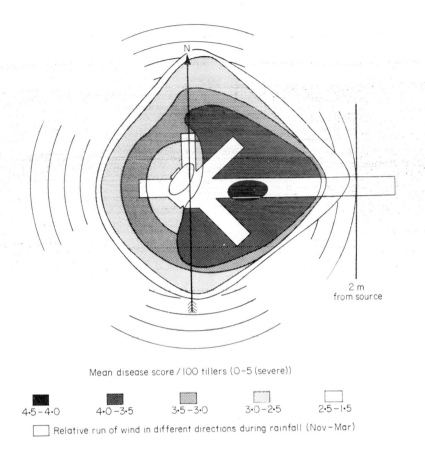

Mean disease score / 100 tillers (0-5 (severe))

| 4·5-4·0 | 4·0-3·5 | 3·5-3·0 | 3·0-2·5 | 2·5-1·5 |

Relative run of wind in different directions during rainfall (Nov-Mar)

Figure 1. Eyespot infection around a point source. The concentric lines represent rows of wheat around the source.

CULTIVAR SUSCEPTIBILITY

Cereal cultivars differ in susceptibility to eyespot infection and it was considered possible that growing resistant cultivars might not only decrease effects of eyespot on crop growth and yield but also affect carry-over of the pathogen. Spore release from samples of the most severely affected culm bases of 12 winter wheat, eight winter barley and four spring barley cultivars grown in 1978, chosen for their range of susceptibilities to the pathogen, was monitored from November 1978 until May 1979. There were no obvious differences between these cultivar samples in lesion severity, and they had similar sporulation periods. Although differing numbers of spores were caught, there was no correlation between numbers of spores released and cultivar susceptibility. Lesions on all cultivars produced numerous spores and differences in carry-over potential between cultivars are unlikely.

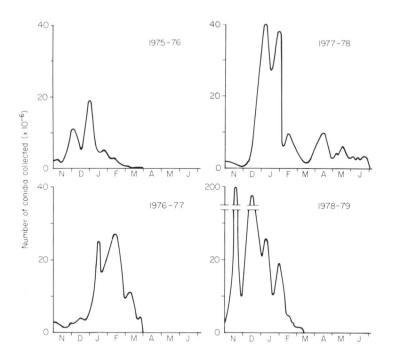

Figure 2. Seasonal fluctuations in numbers of *Pseudocercosporella herpotrichoides* conidia collected in rain-washings from 15 infected culm bases.

INFECTION CONDITIONS AND TIME OF INFECTION

Knowing when spores are produced may indicate when disease is likely to develop but only spores capable of infecting are important. Frequency of eyespot infection periods was monitored during the 1977 and 1978 cropping seasons by exposing healthy wheat and barley to eyespot inoculum dispersed during periods of rain, changing the test plants after each period of rain. Symptoms developed after 87% of rain occasions but only when more than 1 mm of rain fell and leaf wetness lasted more than 6 h.

In 1975/76, plots of winter wheat cv. Maris Huntsman were inoculated at weekly intervals from 28 October (GS 14) until 3 May (GS 37). By 26 May, plots inoculated between 15 December and 5 January were most severely affected, whereas symptoms were only just appearing in plots inoculated after 9 February. By harvest, symptoms were present in all plots, but severe lesions occurred only in plots inoculated before 30 March. Again, plots inoculated between 15 December and 5 January were most severely affected.

FUNGICIDE ACTIVITY

An examination of the way fungicides work has shown that most of those recommended for eyespot control in the United Kingdom suppress disease development by decreasing fungal colonization of the stem but more fungicide needs to be there than is usually provided by a normal spray.

In glasshouse and small plot experiments, carbendazim applied up to 56 days after initial infection arrested disease progress but complete control was obtained only when carbendazim was applied before infection. In one test, carbendazim-treated wheat and barley, subsequently exposed to spores of *P. herpotrichoides* from infected culm bases for 56 days, remained healthy until harvest. There was no significant difference between the curative activities of benomyl, carbendazim, carbendazim plus maneb and thiophanate-methyl, but carbendazim was the best protectant.

Thus more effective control of eyespot might be achieved if carbendazim were applied earlier, and not later than 56 days after the first infection period. Sprays at this time should arrest all existing infection and protect the crop for a further 56 days. The effectiveness of such early sprays was tested on winter wheat cv. Maris Huntsman, grown in 1977/8, on a site with much eyespot-infected debris remaining from the previous wheat crop. Carbendazim application on 23 December, one day after much inoculum had been released from the infected stubble, was compared with the recommended spring application at GS 32. Prior to the spring application, 31 to 35% of culms were infected in unsprayed plots whereas in plots that had received carbendazim in December only 5% were infected. At harvest, winter and spring applications gave similar decreases in disease and increases in yield. However, in this year spores were released from debris until June and with both winter and spring applications it is unlikely that sufficient fungicide remained to protect plants throughout the entire period of risk.

Combinations of winter and spring sprays were also tested in 1978/9. At GS 32, 50% of culms were infected in unsprayed plots whereas only 7% of culms were infected in plots sprayed with carbendazim in December. In this year spore release from debris ceased in mid-March but, even so, insufficient carbendazim apparently remained in plants treated in December because eyespot lesions had developed on them by July. The spring application alone significantly decreased both the number of infected culms and lesion severity but eyespot was only completely controlled in plots sprayed in both winter and spring. Severe eyespot lesions had developed in unsprayed plots by July but the crop did not lodge, and yield was not significantly increased by any of the fungicide treatments.

CONCLUSIONS

The duration of sporulation from infected stubble determines which crops are at risk and for how long. Thus in most years winter cereals are at risk from November until March, whereas spring cereals emerging

from late March to early April are less exposed to inoculum. In many
crops most infection occurs before the recommended time for applying
fungicides. Therefore the usual, recommended time for a single spray is
best if the fungicide is curative but not if it is only protectant.

Sprays to control eyespot are recommended if 15-20% of plants are
infected at the late tillering stages (GS 31) but this assumes accurate
diagnosis. Furthermore, some infected culms may not show symptoms at
this growth stage, depending on when they were infected and the length
of the latent period. It may be more appropriate to make spray
decisions on the basis of previous cropping, supplemented by regional
information on spore release and the suitability of weather for
infection.

Long-term strategies should also be considered. The use of autumn
and spring sprays in combination may eliminate eyespot from a crop and
so prevent carry-over on stubble. However, it is not known how long
fields treated in this way would remain free of the pathogen.

REFERENCES

Bruehl G.W. & Cunfer B. (1972) Control of Cercosporella foot rot of
    wheat by benomyl. *Plant Disease Reporter* 56, 20-3.
Fehrmann H. & Schrödter H. (1973) Control of *Cercosporella herpotri-
    choides* in winter wheat in Germany. *Proceedings 7th British
    Insecticide and Fungicide Conference* 1, 119-26.
Glynne M.D. (1953) Production of spores by *Cercosporella herpotri-
    choides*. *Transactions of the British Mycological Society* 36, 46-51.
Hampel M. & Löcher F. (1973) Control of cereal diseases with carbend-
    azim. *Proceedings 7th British Insecticide and Fungicide Conference*
    1, 127-34.
Huber D.M. & Mulanax M.W. (1972) Benomyl rates and application time
    for wheat foot rot control. *Plant Disease Reporter* 56, 342-4.
Jordan V.W.L & Tarr H.S. (1979) Eyespot (*Pseudocercosporella herpo-
    trichoides*):effect on yield. *Report of the Long Ashton Research
    Station for 1978*, pp. 143-4.
Ponchet J. (1959) La maladie du piétin-verse des céréales; *Cercospor-
    ella herpotrichoides* Fron. Importance agronomique, biologie,
    epiphytologie. *Annales des Épiphyties* 10, 45-98.
Powelson R.L. & Cook G.E. (1969) *Fungicide and Nematicide Tests* 25, 97.
Rule J.S. (1975) The effect of benomyl on *Cercosporella herpotrichoides*
    and yield in winter wheat. *Proceedings 8th British Insecticide and
    Fungicide Conference* 2, 397-403.
Scott P.R. & Hollins T.W. (1974) Effects of eyespot on the yield of
    winter wheat. *Annals of Applied Biology* 78, 269-79.
Slope D.B., Etheridge J. & Callwood J.E. (1970) The effect of flame
    cultivation on eyespot disease of winter wheat. *Plant Pathology* 19,
    167-8.
Taylor J.S. & Waterhouse S. (1975) The use of carbendazim in the
    United Kingdom for the control of *Cercosporella herpotrichoides* in
    winter wheat. *Proceedings 8th British Insecticide and Fungicide*

Conference <u>2</u>, 389-96.
Zadoks J.C., Chang T.T. & Konzak C.F. (1974) A decimal code for the
  growth stages of cereals. *Weed Research* <u>14</u>, 415-21.

Leaf surface micro-organisms as pathogen antagonists and as minor pathogens

C. H. DICKINSON

Department of Plant Biology, The University,
Newcastle-upon-Tyne

INTRODUCTION

Large numbers of bacteria and fungi can occur on cereal leaves, though
these populations are not as luxuriant as those on longer-lived leaves
(Dickinson, 1976). However knowledge of the species composition of
cereal phylloplane populations is incomplete. For example, practically
nothing is known about which bacteria occur on cereal leaf surfaces,
though studies of *Lolium perenne* suggest that they probably include Gram-
negative genera such as *Erwinia*, *Flexibacter*, *Pseudomonas* and *Xanthomonas*, Gram-positive genera such as *Micrococcus* and *Staphylococcus*, and
Gram-variable genera such as *Listeria* (Austin *et al.*, 1978). More is
known about the yeasts on cereal leaves and heads. Ballistospore-
forming genera of the Basidiomycotina (*Sporobolomyces* and *Tilletiopsis*)
are often present in large numbers (Last, 1955; Fokkema & van der Meulen,
1976; Dickinson & Wallace, 1976) and these may be accompanied by species
of *Cryptococcus*, *Rhodotorula*, *Torulopsis* and *Trichosporon* (Flannigan &
Campbell, 1977). Many filamentous fungi have also been isolated from
the surfaces of healthy, green, cereal leaves (Dickinson, 1976) but it
is not known which species grow well in this habitat. The available
evidence suggests that relatively few species make substantial growth on
these aerial plant surfaces, but those that do include *Alternaria
alternata*, *Aureobasidium pullulans*, *Cladosporium cladosporioides* and
C. herbarum.

The effects of environmental conditions on phylloplane populations
are only poorly understood but exogenous nutrients, such as the con-
stituents of the cuticle, exudates from within the leaf, aphid honeydew,
pollen, microbial propagules, and wind-blown and splash-dispersed debris,
are necessary for the growth of many yeasts (Bashi & Fokkema, 1977). By
contrast, some filamentous fungi can grow extensively in the absence of
exogenous nutrients (Dickinson & Bottomley, 1980). Amongst the physical
factors, water and rapid and frequent changes in leaf surface temperature are probably critical. The nature of this habitat suggests that
the microbial inhabitants must be able to alternate rapidly between
active growth and mere survival.

Jenkyn J. F. & Plumb R. T. (1981) *Strategies for the Control of Cereal Disease*

INFLUENCE OF FOLIAR FUNGICIDES ON THE PHYLLOPLANE MICROFLORA

Some of the foliar fungicides approved for use on cereals are very specific and do not interfere with the growth of the non-parasitic microbes inhabiting leaf surfaces. By contrast other chemicals, including captafol, carbendazim-generating compounds and dithiocarbamates, have broad spectra of activity against both pathogenic and non-pathogenic organisms (Dickinson, 1973; Jenkyn & Prew, 1973; Dickinson & Wallace, 1976; Mappes & Hampel, 1977; Andrews & Kennerley, 1978).

It seems probable that any chemical with a broad spectrum of activity will have a considerable impact on any ecosystem in which it is used. The effects of such chemicals are also likely to be enhanced by the current trend towards formulations containing mixtures of fungicides which increases their spectra of activity against not only pathogens but also non-pathogens. Furthermore, many cereal fungicides have been cleared for use during most of the crop's life; in some instances from germination to seven days before harvest, so that large disturbances in the epiphytic microflora can be caused at almost any time. However, we know so little about the incidental recipients of fungicides that it is hard to monitor the effects of the compounds on individual microbial groups, still less their broader ecological impact.

Although the role of epiphytic micro-organisms on cereal leaves is not well understood it is possible that non-pathogenic organisms on aerial plant surfaces interact with pathogens. Such interactions may restrict pathogen development, a form of biological control, or enhance pathogen success. Senescence of leaves, flowers and fruits is accompanied, and possibly influenced, by prolific microbial activity, mainly of filamentous fungi. These effects can obviously be influenced by broad spectrum fungicides.

INTERACTIONS BETWEEN SAPROPHYTIC AND PATHOGENIC ORGANISMS ON AERIAL PLANT SURFACES

In recent years environmental and economic considerations have been powerful stimuli encouraging the development of biological systems of disease control. Much of the early work concerned root pathogens (Baker & Cook, 1974) but some additional examples of antagonistic interactions on aerial plant surfaces were documented in Dickinson & Preece (1976).

In most considerations of biological control, attention is focused on specialized hyperparasites which attack particular pathogens. Such organisms include *Cicinnobolus* spp. and *Tilletiopsis* spp., which parasitize the powdery mildews, and *Eudarluca* spp. and *Tuberculina* spp., which attack the rusts. However, it is not even known if these organisms are sufficiently common for them to be a useful natural check on pathogen populations and we certainly have insufficient information to enable us to usefully manipulate them.

Significant effects on pathogens have been achieved using non-parasitic members of the phylloplane microflora (Table 1). Some of these examples involve yeasts and bacteria which, as noted earlier, are capable of prolific growth on leaf surfaces. Others involve filamentous fungi which are not noted for their extensive epiphytic growth. However in most of these latter examples spores of the antagonist were inoculated onto leaves together with those of the pathogen. Hence, if the environmental conditions favoured germination of the pathogen's spores, they probably also favoured the germination of the saprophyte's so permitting interactions between germ tubes. Control of biotrophic pathogens, such as the rust fungi, depends on the prior inoculation of antagonists onto leaves. This may not be so necessary for the control of necrotrophic pathogens, which frequently have a longer vulnerable period of leaf surface activity.

Table 1. Examples of interactions between cereal foliar and head pathogens and other phylloplane organisms.

Pathogen	Antagonist	Reference
Cochliobolus miyeabeanus	*Candida*	Akai & Kuramoto (1968)
Cochliobolus sativus	*Aureobasidium, Cladosporium*	Diem (1969)
Cochliobolus sativus	*Cryptococcus, Sporobolomyces*	Fokkema, Kastelein & Post (1979)
Leptosphaeria nodorum	*Botrytis, Cladosporium, Stemphylium*	Dickinson & Skidmore (1976)
Leptosphaeria nodorum	*Aureobasidium, Cryptococcus, Sporobolomyces*	Fokkema & van der Meulen (1976)
Leptosphaeria nodorum and *Fusarium* spp.	*Cladosporium* and other saprophytes	Zwatz (1976)
Puccinia graminis	Several filamentous fungi	Mishra & Tewari (1976)
Puccinia pennisetti	*Aspergillus, Chaetomium, Fusarium*	Kapooria & Sinha (1969)
Puccinia striiformis	*Alternaria*	Lloyd (1969)
Puccinia spp.	*Bacillus*	Morgan (1963)
Puccinia spp.	*Xanthomonas*	Pon *et al.* (1954)

Most of the interactions listed in Table 1 were identified in experiments where natural populations of antagonists were increased by applying to the leaves inoculum or nutrients known to enhance the antagonists' activity. Recent experiments by Fokkema, den Houter, Kosterman & Nelis (1979) illustrate both of these techniques. Inocula of two yeasts were applied to wheat leaves, together with nutrients known to enhance yeast growth. The resulting enhanced phylloplane flora restricted infection by both *Cochliobolus sativus* and *Leptosphaeria (Septoria) nodorum*. The treatments had some effect throughout the season but the difference between treated and untreated leaves decreased towards harvest. The natural phylloplane flora eventually reached inhibitory levels, or became more active, albeit at a rather late date.

There are few other examples of naturally occurring biological control which are as convincing as that reported by Fokkema, den Houter, Kosterman & Nelis (1979) but this may reflect the small number of investigations reported to date. The application of broad spectrum fungicides will clearly interfere with the build-up of populations of potential antagonists and hence delay the inception of natural biological control mechanisms.

The converse possibility is that epiphytic organisms are harmful and have additive or synergistic effects in combination with recognized foliar pathogens. Synergism appears to be a rare phenomenon in plant pathology (Dickinson, 1979) but some examples of enhanced pathogenic activity have been reported. The chlorotic symptoms caused by *L. nodorum* and *Septoria tritici* on wheat leaves were increased in the presence of *Sporobolomyces* and other saprophytes (Schild & Harrower, 1978). *Botrytis squamosa* was stimulated by several bacteria isolated from onion leaves (Clark & Lorbeer, 1977), and a similar result was obtained with *Alternaria brassicicola* and bacteria from cabbage leaves (H.T. Al-Hadithi, personal communication). More examples of this type of interaction will no doubt be reported when greater attention is directed towards finding them.

Thus, there is evidence that many interactions occur between foliar pathogens and epiphytic saprophytes. The significance of these interactions is not yet known as no surveys have investigated their occurrence and little attention has been paid to crops where disease was slight or absent. It may be over-optimistic to rely on phylloplane microbes for effective biological control of all foliar pathogens but equally it would be foolish for us to allow this flora to be destroyed by toxic sprays before determining its significance.

MICROBIAL ACTIVITY DURING SENESCENCE AND ITS EFFECTS ON GRAIN YIELD AND QUALITY

Applications of broad spectrum fungicides late in the growth of a cereal crop can have marked effects on phylloplane fungi. These effects can often be clearly seen as the ripening flag leaves and heads remain free of sooty moulds. Such filamentous, dematiaceous fungi normally become

particularly active during senescence and their growth can result in
considerable discolouration of the ripe shoots.   There is, however, some
doubt as to the precise identity of the fungi involved.   A survey, in
1976, of crops from various parts of the United Kingdom (UK) showed that
Cladosporium spp. were the most abundant sooty moulds on winter wheat.
This was confirmed by results from North Yorkshire in 1979 (Table 2).
Cladosporium spp. were frequently accompanied by *Alternaria alternata*
and more rarely by *Epicoccum purpurascens*.   However, hyaline Hyphomy-
cetes, such as *Fusarium* spp. and *Verticillium lecanii* have also been
reported to be common on ripe cereal heads (Flannigan & Campbell, 1977).

Table 2.   Microscopic assessments of senescence and fungi on wheat
heads in North Yorkshire, 1979.   Assessments are the means for 15
heads/cultivar for each sampling date.   Assessments were made of the
extent of head senescence, as a percentage, and the colonization of
glumes by fungal hyphae and conidiophores, as percentage area
covered.

| | Cultivar | | | | | |
	Aquila	Bounty	Bouquet	Hobbit	Maris Huntsman	Kador
8 August						
Growth stage*	87	84	85	84	87	88
Senescence	55	30	40	40	40	65
Cladosporium	1.4	0.9	0.7	0.4	1.4	1.0
Erysiphe	2.5	4.2	4.8	1.5	0.7	3.1
22 August						
Growth stage*	91	91	90	91	91	92
Senescence	100	98	95	95	100	100
Cladosporium	20.8	12.5	19.7	22.1	18.8	31.7
Alternaria	6.3	5.7	7.1	7.8	10.3	6.1
Erysiphe	1.4	2.5	2.7	0.3	0.2	2.0

* Growth stage according to Zadoks *et al.* (1974)

Early attempts to correlate the rate of senescence of leaf tissues
with the activity of phylloplane fungi used detached leaves and, though
there are now doubts about the wisdom of using excised tissues, leaves
inoculated with *A. alternata* and *Stemphylium botryosum* senesced faster
than uninoculated leaves (Skidmore & Dickinson, 1973).   Recent experiments have used whole plants but interpretation of the results obtained
is sometimes difficult.   In several instances populations of phylloplane
micro-organisms on senescing tissues have been monitored using cultural
techniques (Mappes & Hampel, 1977; Fokkema, Kastelein & Post, 1979).

Although these techniques are suitable for assessing populations of bacteria and yeasts, they are not satisfactory for filamentous fungi, which probably have the greatest effect on senescence.

Measuring senescence in field and experimental crops is also diffi-cult. The commonest technique involves measurement of leaf chlorophyll but, despite its obvious simplicity and relevance, the results have frequently failed to correlate with comparative visual assessments of experimental plots. These discrepancies may have been due to the difficulty of processing sufficiently large samples to represent the many varied physiological conditions which exist in different leaves at a late stage in a crop's development, and by the absence of any stable characteristic to which to relate the amounts of chlorophyll. Leaf weight has frequently been used for this despite the fact that during senescence leaves lose water and metabolites. Enzyme and total protein content are equally unsatisfactory because they also alter during senescence. Changes in fine structure do give a clear indication of senescence but cannot be readily determined for large numbers of samples. The supply of carbohydrate to the grain might be a better indicator of differences in senescence between various treatments.

Difficulties in interpreting experiments on senescence may also arise if fungicides are used to inhibit phylloplane activity. Although, with care, combinations of fungicide treatments can be designed to counter most obvious criticisms, their use does complicate experiments and it is important to monitor their effects on recognized pathogens and to realize that they can have direct effects on plant development.

Nevertheless, several attempts have been made to assess the effects of phylloplane fungi on crop senescence and yield. In field trials in 1973 and 1974, phylloplane populations were effectively limited by fun-gicides (Table 3) (Dickinson & Wallace, 1976). In both years control of phylloplane saprophytes was correlated with significant yield increases (Dickinson & Walpole, 1975) and plots with the best yielding treatments were observed to stay green longer and contained most chlorophyll. These data do not allow us to relate cause and effect precisely but do provide circumstantial evidence that the phylloplane microflora accelerates crop senescence.

Fokkema, Kastelein & Post (1979) failed to repeat our results but a direct comparison of their experiment with ours is impossible as they used plate counts to quantify the leaf microflora. We also failed to delay senescence or obtain a yield increase in a similar experiment done in 1978 (Table 3) but microbiological analyses showed that there had been less microbial development on the unsprayed shoots than in 1973 and 1974. In addition, the fungicides were relatively ineffective in con-trolling the phylloplane microflora in 1978.

Mappes & Hampel (1977) also obtained evidence that control of the cereal phylloplane flora can result in significant yield increases and there are also many reports of unexplained yield increases resulting from fungicide treatments (Jenkins et al., 1972; Formigoni & Antonelli,

1974; Fehrmann *et al.*, 1978).

Table 3. Effects of fungicides on wheat flag leaf phylloplane populations (numbers of yeast cells and total hyphal length (μm) per mm^2 of leaf surface) and yield (t/ha).

	1973 U*	F*	1974 U	F	1978 U	F
Yeast cells	4272	412	2903	1119	1022	928
Filamentous fungi	4593	286	3136	1054	2700	1767
Yield	4.56	5.16	6.20	7.62	6.35	6.33

* U, untreated; F, fungicides. Those used in 1973, 1974 and 1978 were Zineb (GS >85), Carbendazim + Zineb (GS 71) and Carbendazim + Maneb (GS >71) respectively.

EFFECTS OF SOOTY MOULDS ON GRAIN QUALITY

Barley grains become contaminated with many fungi between growth stages (GS; Zadoks *et al.*, 1974) 75 and 85 (Warnock, 1973). *Alternaria* spp. and *Cladosporium* spp. seem to come from dying flower parts but *Aspergillus* and *Penicillium* originate from the soil (Flannigan, 1978).

If sufficiently numerous, these fungi decrease grain quality. Blackening of the grain makes it unsuitable for flour (Baker *et al.*, 1958) and contamination by some fungi can lead to problems of toxicity when it is fed to animals. Severe contamination may affect seed quality (Bennett, 1928; Dorovskaya & Khasanova, 1974).

PATHOGENICITY OF ALTERNARIA AND CLADOSPORIUM SPECIES

Some of the yield increases discussed above might also be explained if the sooty moulds were parasitic or pathogenic on cereals. Various investigations have shown that both *A. alternata* and *Cladosporium* spp. can be pathogenic on cereals, although they usually need extremely favourable conditions for successful establishment in host tissues.

Field records of A. alternata *and* Cladosporium spp. *as cereal pathogens*

A. alternata has been reported to cause barley and wheat root rots (Ghobrial, 1975; Ermekova & Rysbaeva, 1978), leaf blotch of barley

(Dhanraj, 1970), black point disease of wheat (Machacek & Greaney, 1938 Huguelet & Kiesling, 1973) and alternariosis of wheat (Korobeinikova, 1975), but it is not generally recognized as a cereal pathogen in the UK.

It is more difficult to assess the pathogenic status of *C. clados-porioides* and *C. herbarum* because of the difficulty in identifying the two species. Mackie (1921) described a particularly severe disease of wheat which he attributed to *C. cladosporioides* and which occurred in a relatively warm region affected by spring and autumn fogs. This report thus supports the view that these fungi are particularly successful pathogens under conditions of continuous high humidity (Dickinson & O'Donnell, 1977). Reports implicating *C. herbarum* as a cereal pathogen refer to it as a foot rot pathogen (Neururer, 1961), as causing a kernel rot of maize (Hoppe, 1964) and as a secondary pathogen on cereals damaged by *Gaeumannomyces graminis* and *Pseudocercosporella herpo-trichoides* (Böning, 1957; Glaeser, 1968).

Our field observations in the UK have shown that *C. cladosporioides* is capable of invading through wounds or senescent tissue and sporul-ating on green wheat leaves. The blackening observed on many heads is mainly due to the production of dark-coloured conidiophores and conidia (Fig. 1, a,b,d). Wefts of superficial mycelium are rarely encountered (Fig. 1c) so that, despite their common name, these fungi behave more like parasites than true epiphytes. The timing of the emergence of these conidiophores also suggests that they are parasites (Table 2). Even if penetration and colonization take only a minimal time the presence of large numbers of conidiophores on partly senescent glumes is evidence that colonization had occurred while the glumes were still green, implying that the fungi are capable of behaving as parasites. On heads the initial infection court may be dying glume tips or the dead anthers (Warnock, 1973).

Evidence implicating A. alternata *and* Cladosporium spp. *as parasites*

After a series of experiments with *C. herbarum*, Bennett (1928) concluded that whilst this species is not pathogenic on wheat, it is "semi-parasitic", establishing itself on dying tissues and hastening their death.

Experimental studies by O'Donnell & Dickinson (1980) have shown that all three species, *A. alternata, C. cladosporioides* and *C. herbarum,* can invade *Phaseolus vulgaris* leaves, though the infections were mostly latent with internal colonization being confined to the substomatal cavities. D. Bottomley (personal communication) has examined the behaviour of these species on wheat leaves and found that most attempted penetrations of green leaves were met with a halo and papilla resistance response (Ride & Pearce, 1979). However, some isolates did become established in green leaves, though internal colonization was again mostly restricted. More extensive colonization occurred during leaf senescence when the fungi invaded the green tissues internally from the dead leaf tips. Hyphae grew basipetally alongside vascular bundles and

Figure 1. a) *Cladosporium cladosporioides* conidiophores emerging from an abaxial glume surface. Some detached conidia can be seen on the glume surface but there is no epiphytic mycelium. b) Rows of *C. cladosporioides* conidiophores emerging from internal colonies in a glume with no epiphytic mycelium. c) Extensive development of *C. cladosporioides* on an abaxial glume surface, with epiphytic mycelium and numerous conidiophores. d) *Alternaria alternata* conidiophores emerging through an abaxial glume surface. Conidia can be seen developing on the conidiophores and lying on the glume surface but there is no epiphytic mycelium.

then extended laterally into adjacent tissues. Invasion of dying leaves resulted in accelerated loss of chlorophyll and an increase in total ribonuclease content compared with uninvaded leaves.

If these fungi are parasitic then strains might be expected to differ in virulence; such differences were demonstrated by O'Donnell & Dickinson (1980). Conversely, cultivars might differ in their resistance. Although little is known about such differences, the data in Table 2 suggest that they do exist.

CONCLUSIONS

Biological control of cereal foliar pathogens has not, to date, been considered a practical proposition. There is, however, some evidence that phylloplane populations can interfere with the development of some necrotrophic pathogens, especially towards the end of the growing season.

Crop senescence seems most affected by a number of filamentous fungi which are ubiquitous on aerial plant tissues. However, their significance in hastening senescence appears to be markedly affected by environmental conditions and possibly by cultivar.

Hence there are at least two, possibly conflicting, considerations concerning the use of fungicides and their possible effects on the phylloplane microflora. Sprays applied late or of persistent compounds may decrease natural biological control but may also control those fung: which would otherwise accelerate senescence. In the future, decisions to spray crops with broad spectrum fungicides may have to take both effects into account.

REFERENCES

Akai S. & Kuramoto T. (1968) Micro-organisms existing on leaves of rice plants and the occurrence of brown leaf spot. *Annals of the Phytopathological Society of Japan* 34, 313-6.
Andrews J.H. & Kennerley C.M. (1978) The effects of a pesticide program on non-target epiphytic microbial populations on apple leaves. *Canadian Journal of Microbiology* 24, 1058-72.
Austin B., Goodfellow M. & Dickinson C.H. (1978) Numerical taxonomy of bacteria isolated from *Lolium perenne*. *Journal of General Microbiology* 104, 139-55.
Baker G.J., Greer E.N., Hinton J.J.C., Jones C.R. & Stevens D.J. (1958) The effect on flour colour of *Cladosporium* growth on wheat. *Cereal Chemistry* 35, 260-75.
Baker K.F. & Cook R.J. (1974) *Biological Control of Plant Pathogens*. W.H. Freeman, San Francisco. 433 pp.
Bashi E. & Fokkema N.J. (1977) Environmental factors limiting growth of *Sporobolomyces roseus*, an antagonist of *Cochliobolus sativus*, on wheat leaves. *Transactions of the British Mycological Society* 68, 17-25.
Bennett F.T. (1928) On *Cladosporium herbarum*: the question of its parasitism, and its relation to "thinning out" and "deaf ears" in wheat. *Annals of Applied Biology* 15, 191-212.
Böning K. (1957) Starkes Auftreten von Schwärze an Getreide. *Pflanzenschutz* 9, 115.
Clark C.A. & Lorbeer J.W. (1977) The role of phyllosphere bacteria in pathogenesis by *Botrytis squamosa* and *B. cinerea* on onion leaves. *Phytopathology* 67, 96-100.
Dhanraj K.S. (1970) Alternaria leaf blotch of barley. *Indian Phytopathology* 23, 116-7.

Dickinson C.H. (1973)   Interactions of fungicides and leaf saprophytes. *Pesticide Science* 4, 563-74.

Dickinson C.H. (1976)   Fungi on the aerial surfaces of higher plants. *Microbiology of Aerial Plant Surfaces* (Ed. by C.H. Dickinson & T.F. Preece), pp. 293-324.   Academic Press, London.

Dickinson C.H. (1979)   External synergisms among organisms inducing disease.   *Plant Pathology - An Advanced Treatise* Vol. IV.   (Ed. by J.G. Horsfall & E.B. Cowling), pp. 97-111.   Academic Press, New York.

Dickinson C.H. & Bottomley D. (1980)   Germination and growth of *Alternaria* and *Cladosporium* in relation to their activity in the phylloplane.   *Transactions of the British Mycological Society* 74, 309-19.

Dickinson C.H. & O'Donnell J. (1977)   Behaviour of phylloplane fungi on *Phaseolus* leaves.   *Transactions of the British Mycological Society* 68, 193-9.

Dickinson C.H. & Preece T.F. (Eds.) (1976)   *Microbiology of Aerial Plant Surfaces*.   Academic Press, London.   669 pp.

Dickinson C.H. & Skidmore A.M. (1976)   Interactions between germinating spores of *Septoria nodorum* and phylloplane fungi.   *Transactions of the British Mycological Society* 66, 45-56.

Dickinson C.H. & Wallace B. (1976)   Effect of late applications of foliar fungicides on activity of micro-organisms on winter wheat flag leaves.   *Transactions of the British Mycological Society* 67, 103-12.

Dickinson C.H. & Walpole P.R. (1975)   The effect of late applications of fungicides on the yield of winter wheat.   *Experimental Husbandry* 29, 23-8.

Diem H.G. (1969)   Micro-organismes de la surface des feuilles.   II. Interactions entre quelques champignons parasites et divers sapro-phytes filamenteux de la phyllosphère de l'orge.   *Bulletin de l'Ecole Nationale Supérieure Agronomique de Nancy* 11, 12-7.

Dorovskaya L.M. & Khasanova G.Sh. (1974)   Prichiny snizheniya polevoi vskhozhesti semya yarovoi pshenitsy v Ural'skoi oblasti.   (Causes of decrease in field germination of spring wheat seeds in the Urals region).   *Vestnik Sel'skokhozyaistvennoi Nauki Kazakhstana* 17, 21-5.

Ermekova B.D. & Rysbaeva K.D. (1978)   Rasprostranenie i vidovoi sostav vozbuditelei kornevykh gnilei.   (Distribution and species composition of causal agents of rootrots).   *Vestnik Sel'skokhozyaistvennoi Nauki Kazakhstana* 21, 45-7.

Fehrmann H., Reinecke P. & Weihofen U. (1978)   Yield increase in winter wheat by unknown effects of MBC-fungicides and captafol.   *Phyto-pathologische Zeitschrift* 93, 359-62.

Flannigan B. (1978)   Primary contamination of barley and wheat grain by storage fungi.   *Transactions of the British Mycological Society* 71, 37-42.

Flannigan B. & Campbell I. (1977)   Pre-harvest mould and yeast floras on the flag leaf, bracts and caryopsis of wheat.   *Transactions of the British Mycological Society* 69, 485-94.

Fokkema N.J., den Houter J.G., Kosterman Y.J.C. & Nelis A.L. (1979)   Manipulation of yeasts on field-grown wheat leaves and their anta-gonistic effect on *Cochliobolus sativus* and *Septoria nodorum*.   *Transactions of the British Mycological Society* 72, 19-29.

Fokkema N.J., Kastelein P. & Post B.J. (1979)   No evidence for acceleration of leaf senescence by phylloplane saprophytes of wheat. *Transactions of the British Mycological Society* 72, 312-5.

Fokkema N.J. & van der Meulen F. (1976) Antagonism of yeastlike phyllosphere fungi against *Septoria nodorum* on wheat leaves. *Netherlands Journal of Plant Pathology* 82, 13-6.

Formigoni A. & Antonelli C. (1974) Due anni di esperienze in Italia con tiofanta metil + maneb nella lotta control le malattie della parte aerea del frumento. *Proceedings of the 4th Conference on Phytiatry and Phytopharmacy in the Mediterranean Region*, pp. 205-14. Ruilliere-Libeccio, Avignon, France.

Ghobrial E. (1975) Fungi causing complex infection of root-rot diseases on barley. *Agricultural Research Review* 53, 15-20.

Glaeser G. (1968) Das Auftreten wichtiger Schadensursachen an Kulturpflanzen in Österreich im Jahre 1967. *Pflanzenschutz-berichte* 37, 67-85.

Hoppe P.E. (1964) Inoculation technique for *Cladosporium* ear rot of corn. *Plant Disease Reporter* 48, 391-3.

Huguelet J.E. & Kiesling R.L. (1973) Influence of inoculum composition on the black point disease of durum wheat. *Phytopathology* 63, 1220-5.

Jenkins J.E.E., Melville S.C. & Jemmett J.L. (1972) The effect of fungicides on leaf diseases and on yield in spring barley in south-west England. *Plant Pathology* 21, 49-58.

Jenkyn J.F. & Prew R.D. (1973) The effect of fungicides on the incidence of *Sporobolomyces* spp. and *Cladosporium* spp. on flag leaves of winter wheat. *Annals of Applied Biology* 75, 253-6.

Kapooria R.G. & Sinha S. (1969) Phylloplane microflora of pearl millet and its influence on the development of *Puccinia pennisetti*. *Transactions of the British Mycological Society* 53, 153-5.

Korobeinikova A.V. (1975) Vredonost' al'ternarioza semjan pshenitsy. (Damage by alternariosis to wheat seeds). *Sashita Rastenii*, 5, 23.

Last F.T. (1955) Seasonal influence of *Sporobolomyces* on cereal leaves. *Transactions of the British Mycological Society* 38, 221-39.

Lloyd E.H. (1969) The influence of low temperature on the behaviour of wheat leaves infected with *Puccinia striiformis* West. *Dissertation Abstracts* 29(9)B, 3156-7.

Machacek J.E. & Greaney F.J. (1938) The "black-point" or "kernel smudge" disease of cereals. *Canadian Journal of Research, Section C* 16, 84-113.

Mackie W.W. (1921) Sooty mold (*Hormodendrum cladosporioides*) on wheat. *Annual Report Agricultural Experimental Station California* 42, 45.

Mappes C.J. & Hampel M. (1977) Yield responses of winter barley to late fungicide treatments. *Proceedings 1977 British Crop Protection Conference - Pests and Diseases* 1, 49-55.

Mishra R.R. & Tewari R.P. (1976) Studies on biological control of *Puccinia graminis tritici*. *Microbiology of Aerial Plant Surfaces* (Ed. by C.H. Dickinson & T.F. Preece) pp. 559-67. Academic Press, London.

Morgan F.L. (1963) Infection inhibition and germ tube lysis of three cereal rusts by *Bacillus pumilus*. *Phytopathology* 53, 1346-8.

Neururer H. (1961) Höhe Ernteverluste durch Füsskrankheiten des Getreides. *Pflanzenarzt* 14, 92-3.

O'Donnell J. & Dickinson C.H. (1980) Pathogenicity of *Alternaria* and *Cladosporium* isolates on *Phaseolus*. *Transactions of the British Mycological Society* 74, 335-42.

Pon D.S., Townsend C.E., Wessman G.E., Schmitt C.G. & Kingsolver C.H. (1954) A *Xanthomonas* parasitic on uredia of cereal rusts. *Phytopathology* 44, 707-10.

Ride J.P. & Pearce R.B. (1979) Lignification and papilla formation at sites of attempted penetration of wheat leaves by non-pathogenic fungi. *Physiological Plant Pathology* 15, 79-92.

Schild D.E. & Harrower K.M. (1978) The colonization of wheat leaves by *Septoria* spp. in the presence of phylloplane saprophytes. *Australian Plant Pathology Society Newsletter* 7, 6-7.

Skidmore A.M. & Dickinson C.H. (1973) Effect of phylloplane fungi on the senescence of excised barley leaves. *Transactions of the British Mycological Society* 60, 107-16.

Warnock D.W. (1973) Origin and development of fungal mycelium in grains of barley before harvest. *Transactions of the British Mycological Society* 61, 49-56.

Zadoks J.C., Chang T.T. & Konzak C.F. (1974) A decimal code for the growth stages of cereals. *Weed Research* 14, 415-21.

Zwatz B. (1976) Getreideschwärze: starkes Auftreten. *Pflanzenarzt* 29, 103-4.

Strategies for avoiding resistance to fungicides

Department of Phytopathology, Agricultural University,
Wageningen, The Netherlands

INTRODUCTION

Fungal plant pathogens may become resistant to chemicals used to
control them, and since the introduction of systemic fungicides 10
years ago, this has become an increasing problem. Such resistance may
develop as a consequence of genetic or non-genetic changes in the
fungal cell. The latter type is usually rapidly lost in the absence
of the fungicide, and is of little practical importance. This chapter
therefore considers only stable, genetically determined resistance.

Fungicide-resistant cells may either appear spontaneously or be
induced by mutagenic agents at a frequency of between 1 in 10^4 and 1 in
10^9. Resistant cells may be present in small numbers, even before the
fungicide is used (Schreiber & Townsend, 1976). While the percentage
of resistant cells is small, there will be no detectable decrease in
fungicide efficiency. Failure of disease control only occurs when,
under the selecting effect of the fungicide, a substantial part of the
pathogen population becomes resistant.

An understanding of the factors which govern the emergence of
resistant cells, and the subsequent development of a predominantly
fungicide-resistant pathogen population, is necessary in order to
develop strategies of fungicide use. The first part of this chapter
considers some of the factors involved in fungicide resistance, and
the second their relevance to particular cereal fungicides.

MECHANISM OF ACTION

The type of fungicide determines whether mutations towards fungicide-
resistance are likely to occur (Dekker, 1976; Georgopoulos, 1977).
Resistance does not readily occur with fungicides that interfere with
fungal metabolism at many sites. Most of our conventional fungicides,
such as the dithiocarbamates and metal toxicants, are of this type.
The occurrence of mutations leading to changes at all sites of action
is unlikely or perhaps impossible, and mutations which change the

Jenkyn J. F. & Plumb R. T. (1981) *Strategies for the Control of Cereal Disease*

permeability of the fungal membrane or enhance detoxification of the chemical do not readily occur.

Resistance to fungicides which act primarily at one specific site in fungal metabolism may require only a single gene mutation. As most, if not all, systemic fungicides are specific-site inhibitors, it is not surprising that, at least in the laboratory, mutants resistant to them can almost invariably be obtained (Van Tuyl, 1977). Indeed one would only expect *not* to detect them when all mutations in the genes governing sensitivity are lethal, although no such examples have been reported.

FITNESS OF RESISTANT CELLS

The detection of resistant mutants in the laboratory does not imply that use of the fungicide will inevitably lead to problems in practice. This depends on the properties of the resistant mutants, such as degree of virulence and sporulation capacity, which determine fitness. Resistant mutants may show much variation, but if individuals occur with a degree of fitness equal to or not much less than that of the sensitive pathogen, a fungicide-resistant population may develop even if selection pressure is only moderate or low. Several such examples have been reported (Bent *et al.*, 1971; Georgopoulos & Dovas, 1973; Nishimura *et al.*, 1973; Ito & Yamaguchi, 1977; Schroth *et al.*, 1979).

However, with some fungicides, such as the antibiotic pimaricin, increased resistance of the pathogen is apparently associated with decreased fitness (Dekker & Gielink, 1979), which should delay or even prevent the build-up of a fungicide-resistant population. For several years pimaricin, only recently introduced for control of bulb rot in narcissus, has been used against fungal pathogens in man and to prevent moulding of freshly made cheese; no problems with resistance have developed. Resistance to some of the new systemic fungicides may also be linked to decreased virulence, as indicated by the work of Fuchs & Drandarevski (1976) with triforine and De Waard & Gieskes (1977) with fenarimol. Although triforine-resistant mutants are easily obtained in the laboratory, no resistance problems in practice have yet been reported. Therefore it seems unlikely that all systemic fungicides will encounter resistance problems.

FUNGICIDE USE

Amount and frequency of application

The amount of fungicide applied, frequency of application and the formulation, especially as it affects persistence, all contribute to selection pressure. Application of a systemic fungicide to soil or seed allows uninterrupted uptake by the plant, and may therefore increase the exposure time.

When highly resistant cells occur, then a high selection pressure will favour the rapid build-up of a resistant population. However, if mutants are only weakly or moderately resistant, then a high selection pressure may prevent their growth and give adequate disease control but a low selection pressure may favour the survival of forms with moderate resistance. To avoid fungicide-resistance, it is therefore important to know what mutants are likely to arise; whether they will have a high level of resistance, as with benomyl, or only a moderate to low level of resistance, as with triforine or pimaricin. However, accumulation of mutations each carrying a low level of resistance may eventually result in forms with a high level of resistance.

Efficiency of treatment and area treated

Thorough treatment provides little opportunity for the sensitive forms of the pathogen to escape, and therefore resistant forms have little competition during periods of relaxed selection pressure. This will favour the build-up of a resistant pathogen population. Similarly if large areas are treated with a single or related chemicals, few sensitive forms will enter the crop from outside.

Alternating or combined use of fungicides

To delay or prevent the development of fungicide resistance, the use of fungicides with different mechanisms of action, either alternately or in combination, has been advocated (Wolfe, 1971). When two site-specific inhibitors are combined the pathogen may acquire resistance to both. It is preferable therefore to use a systemic fungicide in combination with a multi-site inhibitor. This is now commonly done, especially where benomyl or other benzimidazole fungicides are used.

A mathematical model to test whether different chemicals should be used in combination (tank mix) or alternately has been developed by Kable & Jeffery (1980). In this model spray coverage had a bigger effect on the rate of selection than the efficacies of the two fungicides against the sensitive and resistant subpopulations. If spray coverage is complete it appears advantageous to use the two fungicides alternately or in a planned sequence. If coverage is less than 99% it is better to use mixtures to decrease the rate of increase of the resistant form. Since complete coverage will seldom be achieved, a mixture appears to be the best strategy. The authors realize that reproduction of the target organism is not included in their model but consider that, provided the subpopulations reproduce at equal rates, it can be ignored. However, in the absence of the fungicide the resistant subpopulation is likely to have a slower reproduction rate than the susceptible. Otherwise, even in the absence of fungicide, the population would gradually shift towards fungicide resistance, which has never happened. Some authors have reported fungicide-resistant forms with equal, or sometimes greater, virulence than the original sensitive form, but it is possible that the resistant strains have not been compared with the most fit sensitive form (Dovas et al., 1976).

Whether two fungicides, A and B, should be used in sequence or in combination, is probably influenced by the relative fitness of the resistant and sensitive forms. Sequential use sllows the pathogen population to shift back in the direction of the A-sensitive form when the use of fungicide A is interrupted and towards the B-sensitive form when the use of B is interrupted. These shifts are enhanced by influx of the sensitive forms from neighbouring fields, and when back mutations occur. Shifts back towards fungicide sensitivity are not possible if a combination of the fungicides is continually present in the plant.

The extent to which resistance problems are delayed by sequential us depends on the speed at which resistant fungal populations build up and the rate of reversion to the fungicide-sensitive forms during interrupt ed selection pressure. If the second fungicide (B) is a multi-site inhibitor, alternation of (A + B) with B might be even more effective i: counteracting the development of resistance.

Negatively correlated cross-resistance

The build-up of fungicide-resistant populations might be totally pre- vented if a mixture of fungicides with negatively correlated cross- -resistance could be used (i.e. resistance to fungicide A is linked wit: sensitivity to fungicide B and *vice versa*). In *Piricularia oryzae* negatively correlated cross-resistance has been reported to occur with phosphoramidate and phosphorothiolate fungicides (Uesugi *et al.*, 1974) and with phosphoramidate and isoprothiolane (Katagiri & Uesugi, 1977). It also occurs with carboxin and antimycin A in one mutant of *Ustilago maydis* (Georgopoulos & Sisler, 1970). In studies with *Aspergillus nidulans*, strains were obtained in which resistance to benomyl was link: with sensitivity to thiabendazole; resistance of other strains to thiabendazole was linked with increased sensitivity to benomyl (Van Tuy. *et al.*, 1974). However, the majority of mutants showed cross-resistanc: with respect to benomyl and thiabendazole. No combination of compounds where negatively correlated cross-resistance is usual is yet used in practice.

TYPE OF DISEASE

Accessibility of the pathogen

Pathogens which are not easily reached by the fungicide, or are in tissues from which the fungicide rapidly disappears, are difficult to eliminate, and escape of sensitive forms then counteracts spread of resistant forms. It is difficult, for example to eliminate vascular wilt pathogens from vessels even with systemic fungicides. Similarly pathogens on roots or in soil escape fungicides applied as sprays. If the fungicide rapidly disappears, then selection pressure is low and favours the sensitive form. This may apply in wheat culms treated with carbendazim to control *Pseudocercosporella herpotrichoides*, because the fungicide is transported acropetally in the transpiration stream.

On the other hand, where the pathogen is difficult to reach, fungicide gradients may form, which might favour survival of strains with moderate to low resistance (Miller & Fletcher, 1974).

Multiplication of the pathogen

Resistance spreads more rapidly if the pathogen sporulates abundantly on aerial plant parts, than if it does not sporulate or has spores which are not easily transported by wind, rain or other carriers. Many soil-borne and root diseases are in the latter category.

Infection threshold

When the infection threshold (i.e. the minimum amount of inoculum necessary for successful infection) is large, the appearance of isolated resistant cells arising by spontaneous mutation, in the presence of the fungicide, rarely results in infection and this decreases the risk of fungicide resistance accumulating in the population.

Life cycle of the pathogen

Interactions between the length of the life cycle and the duration of selection pressure have been discussed by Wolfe (1975). He suggests that, if the life cycle is longer than the period that the miminal effective dose is present, a considerable selective advantage for resistance may exist, since the selected organism will presumably be well-fitted to environments both with and without fungicide. If the life cycle is shorter, individuals selected for resistance may be unsuited to the environment after the disappearance of the fungicide.

The ability of the pathogen to transmit resistance via heterokaryosis (fungi) or via plasmids (certain bacteria) might contribute to the speed of build-up of a resistant population.

FUNGICIDE RESISTANCE IN CEREAL PATHOGENS

Until the advent of systemic compounds, the use of fungicides on cereals in Western Europe was restricted to seed disinfection, especially with organic mercury compounds. As previously discussed, rapid development of resistance to multi-site inhibitors, such as organic mercury compounds, is unlikely to occur but in at least one case prolonged use has resulted in resistance (*Pyrenophora avenae* on oats, Noble *et al.*, 1966); the resistance mechanism is not known.

Recently several systemic fungicides have been introduced for the control of various cereal diseases. Among these are benzimidazole derivatives such as benomyl, and the pyrimidine derivative ethirimol, to both of which resistance has been detected.

Benzimidazole fungicides against eyespot and glume blotch

The benzimidazole fungicides (e.g. benomyl and carbendazim) and the carbendazim precursor thiophanate-methyl are very effective against eyespot. However, because development of resistance to these fungicide has been reported for various pathogens it was feared that continual us against eyespot might lead to resistance problems. Horsten (1979) treated field-grown winter barley with one or two sprays of carbendazim per season. At the milky ripe stage the pathogen was isolated from eyespot lesions, multiplied on wheatmeal agar and spore suspensions the plated out on agar containing 3 µg/ml of carbendazim. The frequency of resistant spores increased (Table 1), but even after three seasons it was still so low that there appeared to be no risk of failure of disease control. Similar results were obtained with winter wheat and rye. Possible reasons for this very slow increase of resistance are:

1. The low selection pressure;
2. The high infection threshold;
3. The slow spread of the pathogen;
4. Survival of sensitive strains, especially on crop debris.

Table 1. Influence of carbendazim applied once per season to winter barley cv. Vogelsanger Gold on the frequency of carbendazim-resistant spores of *Pseudocercosporella herpotrichoides* (Horsten, 1979).

Fungicide treatment	Number of isolates	Number of spores plated out (x 10^{-6})	Frequency of resistant colonies (x 10^9)	% of isolates with resistant spores
1975				
unsprayed	40	8.6	1.28	12
carbendazim	40	7.9	1.27	17
1976				
unsprayed	34	16.2	1.43	12
carbendazim	46	23.5	1.72	46
1977				
unsprayed	19	5.1	4.00	21
carbendazim	18	4.7	8.33	67

Horsten (1979) considers that the use of benzimidazole fungicides against eyespot, at current rates, will not lead to resistance problems. However, once resistance is detectable, only a few more sprays are needed for resistance to dominate (Kable & Jeffery, 1979).

The effects of thiophanate-methyl and of edifenphos against *Septoria nodorum* on wheat were also studied in the field (Horsten, 1979). In unsprayed plots weakly carbendazim-resistant strains were present at a frequency of 1 in 7×10^6 and highly resistant strains were 100 times less common. After 5 years of treatment with thiophanate-methyl there was a significant increase in the frequency of weakly carbendazim-resistant spores, but there were few in total even with four applications per season. There was no clear increase in the frequency of highly resistant spores. In climate chambers, exposure of *S. nodorum* to thiophanate-methyl or edifenphos for nine infection cycles led to a gradual decrease in control (Horsten, 1979). This change was less when the fungicides were used alternately or in combination.

Ethirimol against powdery mildew on barley

Ethirimol, a pyrimidine derivative, has been used for control of barley powdery mildew (*Erysiphe graminis*) since 1969, both as a seed treatment and as a foliar spray. Although resistant isolates can be obtained from ethirimol-treated crops (Wolfe, 1971) and surveys have indicated a slight decrease in sensitivity in these crops, the chemical has continued to be effective and has given yield increases for a number of years (Shephard et al., 1975). These authors report oscillations in sensitivity during the season. Where the fungicide is used strains with intermediate sensitivity are favoured. Although strains with high resistance can be isolated, they tend not to increase in the presence of ethirimol, because they are apparently less competitive (Shephard et al., 1975). Strains of intermediate sensitivity to ethirimol may exert a stabilizing effect within natural populations, so selection pressure should not be increased, as it may improve the relative fitness of insensitive forms (Hollomon, 1978). The selection pressure exerted by ethirimol in the field is not continuously high, because it is usually used only once per season on only one quarter of the crop and is rapidly broken down in plant tissue (Wolfe & Dinoor, 1973).

Kasugamycin against blast

The antibiotic kasugamycin, used to control rice blast (*Piricularia oryzae*), has become less effective since 1971, due to the development of a kasugamycin-resistant pathogen population. This has been attributed to the frequent and almost exclusive use of this antibiotic in certain areas (Miura et al., 1975). After use of kasugamycin was stopped during 1973, the percentage of resistant individuals rapidly decreased (Misato, 1975), suggesting that kasugamycin-resistant strains are less competitive than sensitive ones. This suggests that kasugamycin could continue to be effective provided that continuous selection pressure is avoided. Indeed no kasugamycin resistance was found in samples collected from

districts where kasugamycin had been used alternately with organophosphorus fungicides or in mixtures with phthalide (Ito & Yamaguchi, 1977).

Other systemic fungicides

Resistance to carboxin has been detected in *Ustilage hordei* (Ben-Yephet
et al., 1975), but the fungicide continues to give effective control of
this disease.

Triforine has been used for several years to control various
diseases, including cereal powdery mildew, without occurrence of
fungicide-resistance problems. In laboratory experiments triforine-
resistant strains of *Cladosporium cucumerinum* have been obtained but
appeared to be less virulent. It has been suggested that resistance to
fungicides that inhibit sterol biosynthesis might always be accompanied
by decreased fitness (Fuchs & Drandarevski, 1976). Similarly tridemorph, also used for some years against powdery mildews, has not yet
encountered resistance problems.

CONCLUSIONS

Reviewing the problems of resistance in insects, Brown (1976) concluded:
"It appears that the best policy is to continue using the insecticide
recommended for the job until the control effect becomes inadequate, and
then replace it with the most insecticidal substitute". This strategy
would be difficult with systemic fungicides, because, apart from
problems of availability, the life span of a fungicide may then become
too short to cover the cost of its development. Ideally ways should be
sought to prolong the useful life of fungicides by delaying or avoiding
development of resistance.

Fungicides to which resistance might be expected should not be used
against diseases which can be controlled adequately by other chemicals
or other control methods. They should be used only against diseases
where:

1. The resistant pathogen population increases only slowly or can
   be controlled by a combination of fungicides and cultural
   methods;
2. Control can be obtained at low selection pressure (one or two
   sprays per season), allowing sensitive forms of the pathogen
   to compete with resistant ones.

The selection pressure can be decreased by:

1. Restricting application of resistance-prone fungicides to
   critical periods;
2. Reducing amounts applied, and frequency of application, to the
   minimum necessary for economic control;
3. Choosing a method of application that minimizes the length of
   exposure of the pathogen to the fungicide;
4. Limiting the area treated with any one fungicide.

The use of a second fungicide, preferably a multi-site inhibitor, may restrict multiplication of resistant forms of the pathogen but it is not recommended that the same combination be used for long. Where the fitness of the resistant form is less than that of the sensitive form it may be preferable to use two chemicals in sequence rather than in a mixture.

Crops should be regularly monitored for the presence of resistant strains. Control methods may then be changed before they fail.

It is important to continue the search for new fungicides to which resistance does not readily develop. Although resistance to conventional fungicides with multi-site inhibition is uncommon, there is no reason why the search for systemic compounds, which are considered to be site-specific inhibitors, should be abandoned. There are great differences in the probability that resistance will develop, even among systemic fungicides. New compounds for which resistance in the pathogen is linked to decreased virulence or death should be sought. Compounds may also be discovered which are not fungicidal, but which increase the resistance of the host. If host resistance could be induced at many sites in the plant, pathogens would be less likely to become resistant.

Finally there are indications that combinations of compounds which show negatively correlated cross-resistance might be developed, thus precluding the emergence of resistant mutants.

REFERENCES

Bent K.J., Cole A.M., Turner J.A.W. & Woolner M. (1971) Resistance of cucumber powdery mildew to dimethirimol. *Proceedings 6th British Insecticide and Fungicide Conference* 1, 274-82.
Ben-Yephet Y., Henis Y. & Dinoor A. (1975) Inheritance of tolerance to carboxin and benomyl in *Ustilago hordei*. *Phytopathology* 65, 563-7.
Brown A.W.A. (1976) How have entomologists dealt with resistance? Symposium on resistance of plant pathogens to chemicals, Kansas City, 13 July. *Proceedings of the American Phytopathological Society* 3, 67-74.
Dekker J. (1976) Acquired resistance to fungicides. *Annual Review of Phytopathology* 14, 405-28.
Dekker J. & Gielink A.J. (1979) Acquired resistance to pimaricin in *Cladosporium cucumerinum* and *Fusarium oxysporum* f.sp. *narcissi* associated with decreased virulence. *Netherlands Journal of Plant Pathology* 85, 67-73.
De Waard M.A. & Gieskes S.A. (1977) Characterization of fenarimol-resistant mutants of *Aspergillus nidulans*. *Netherlands Journal of Plant Pathology* 83, *Supl. 1*, 177-88.
Dovas C., Skylakakis G. & Georgopoulos S.G. (1976) The adaptability of the benomyl resistant population of *Cercosporella beticola* in Northern Greece. *Phytopathology* 66, 1452-6.
Fuchs A. & Drandarevski C.A. (1976) The likelihood of development of

resistance to systemic fungicides which inhibit ergosterol biosynthesis. *Netherlands Journal of Plant Pathology* 82, 85-7.

Georgopoulos S.G. (1977) Pathogens become resistant to chemicals.
Plant Disease, An Advanced Treatise (Ed by J.G. Horsfall & E.B.
Cowling), vol. 1, pp. 327-45. Academic Press, New York.

Georgopoulos S.G. & Dovas C. (1973) A serious outbreak of strains of
Cercosporella beticola resistant to benzimidazole fungicides in
Northern Greece. *Plant Disease Reporter* 57, 321-4.

Georgopoulos S.G. & Sisler H.D. (1970) Gene mutation eliminating antimycin A-tolerant electron transport in *Ustilago maydis*. *Journal of
Bacteriology* 103, 745-50.

Hollomon D.W. (1978) Competitive ability and ethirimol sensitivity in
strains of barley powdery mildew. *Annals of Applied Biology* 90,
195-204.

Horsten J.A.H.M. (1979) Acquired resistance to systemic fungicides of
Septoria nodorum and *Cercosporella herpotrichoides* in cereals.
Dissertation, Agricultural University, Wageningen. 107 pp.

Ito I. & Yamaguchi T. (1977) Occurrence of kasagamycin resistant rice
blast fungus, influenced by the application of fungicides. *Annals
of the Phytopathological Society of Japan* 43, 301-3.

Kable P.F. & Jeffery H. (1980) Selection for tolerance in organisms
exposed to sprays of biocide mixtures: a theoretical model. *Phytopathology* 69, 8-12.

Katagiri M. & Uesugi Y. (1977) Similarities between the fungicidal
action of isoprothiolane and organophosphorus thiolate fungicides.
Phytopathology 67, 1415-7.

Miller M.W. & Fletcher J.T. (1974) Benomyl tolerance in *Botrytis
cinerea* isolates from glasshouse crops. *Transactions of the British
Mycological Society* 62, 99-103.

Misato T. (1975) The development of agricultural antibiotics in Japan.
*Proceedings of the First Intersectional Congress of the International
Association of Microbiological Societies 1-7 Sept. 1974, Tokyo* 3,
589-97.

Miura H., Ito H. & Takahashi S. (1975) Occurrence of resistant strains
of *Pyricularia oryzae* to rice blast. *Annals of the Phytopathological
Society of Japan* 41, 415-7.

Nishimura S., Kohmoto K. & Udagawa H. (1973) Field emergence of
fungicide-tolerant strains in *Alternaria kikuchiana* Taneka. *Report
of the Tottori Mycological Institute (Japan)* 10, 677-86.

Noble M., Macgarvie Q.D.., Hams A.F. & Leafe E.L. (1966) Resistance
to mercury of *Pyrenophora avenae* in Scottish seed oats. *Plant
Pathology* 15, 23-8.

Schreiber L.R. & Townsend A.M. (1976) Naturally occurring tolerance in
isolates of *Ceratocystis ulmi* to methyl-2-benzimidazole carbamate
hydrochloride. *Phytopathology* 66, 225-7.

Schroth M.N., Thomson S.V. & Moller W.J. (1979) Streptomycin resistance
in *Erwinia amylovora*. *Phytopathology* 69, 565-8.

Shephard M.C., Bent K.J., Woolner M. & Cole A.M. (1975) Sensitivity to
ethirimol of powdery mildew from UK barley crops. *Proceedings 8th
British Insecticide and Fungicide Conference* 1, 59-66.

Uesugi Y., Katagiri M. & Noda O. (1974) Negatively correlated cross-
resistance and synergism between phosphoramidates and phosphorothio-

lates in their fungicidal actions on rice blast fungi. *Agricultural and Biological Chemistry* 38, 907-12.

Van Tuyl J.M. (1977) Genetics of fungal resistance to systemic fungicides. *Mededelingen Landbouwhogeschool Wageningen* 77, 1-136.

Van Tuyl J.M., Davidse L.C. & Dekker J. (1974) Lack of cross resistance to benomyl and thiabendazole in some strains of *Aspergillus nidulans*. *Netherlands Journal of Plant Pathology* 80, 165-8.

Wolfe M.S. (1971) Fungicides and the fungus population problem. *Proceedings 6th British Insecticide and Fungicide Conference* 3, 724-34.

Wolfe M.S. (1975) Pathogen response to fungicide use. *Proceedings 8th British Insecticide and Fungicide Conference* 3, 813-22.

Wolfe M.S. & Dinoor A. (1973) The problems of fungicide tolerance in the field. *Proceedings 7th British Insecticide and Fungicide Conference* 1, 11-9.

Chemicals in the control of cereal virus diseases

R. T. PLUMB

Rothamsted Experimental Station,
Harpenden, Hertfordshire

INTRODUCTION

Although in 10 or 20 years chemicals used to control virus diseases may include those that have a direct effect upon the virus (Kassanis & White, 1977), or modify the behaviour and biology of the virus vector (Lewis, 1977), we are, at present, concerned only with those that kill the vectors. Although many such pesticides are available their use is not necessarily economic or wise. This chapter considers those factors that influence the incidence of virus diseases of cereals in Britain, and in what circumstances, and with what effects, chemicals might be used to control them.

VIRUSES INFECTING CEREALS IN BRITAIN

More than 50 viruses and mycoplasmas are known to infect cereals (Slykhuis, 1976), and nine have been reported from Britain (Table 1).

Only once has natural infection of wheat with agropyron mosaic virus (AgMV) (Plumb & Lennon, 1980) or oats with ryegrass mosaic virus (RMV) (Mulligan, 1959) been reported. Both viruses are transmitted by the eriophyid mite *Abacarus hystrix* (Nal.) and infection of cereals by them was probably associated with heavy mite infestations of infected *Agropyron repens* (L.) Beauv. and *Lolium* spp. respectively, present as weeds in the cereal crops. Unless infected grass hosts are widespread in cereals it seems unlikely that AgMV or RMV will become important, even locally.

Young wheat is very susceptible to, and severely damaged by, cocksfoot mottle virus (CFMV) when inoculated with infective sap but, in crops, the virus can only be transmitted to wheat by beetles (*Oulema* spp.). Beetles are inefficient vectors and generally invade wheat when it is no longer susceptible (Benigno & A'Brook, 1972), so infection is rare. However, oat sterile dwarf virus (OSDV) and European wheat striate mosaic (EWSM), the latter of unknown aetiology, are quite commonly found in cereals. Both are transmitted by the plant

Jenkyn J. F. & Plumb R. T. (1981) *Strategies for the Control of Cereal Disease*

Table 1. Viruses* reported from cereal crops in the United Kingdom

Virus	Vectors	Transmission by sap	Occurrence	Reported cereal host
Agropyron mosaic virus	Mite	+	Rare	Wheat
Ryegrass mosaic virus	Mite	+	Rare	Oats
Cocksfoot mottle virus	Beetles	+	Rare	Wheat
Oat sterile dwarf virus	Hoppers	–	Rare	Oats
European wheat striate mosaic	Hoppers	–	Widespread but infrequent	Wheat
Oat mosaic virus	Fungus	+	Locally very damaging	Oats
Oat golden stripe virus	Fungus	+	Locally very damaging	Oats
Barley yellow mosaic virus	Fungus	+	Unknown	Barley
Barley yellow dwarf virus	Aphids	–	Ubiquitous	Wheat Barley Oats

* None of the viruses is seed-borne.

hopper *Javesella pellucida* (F.), which, when collected from Lolium-dominated swards, frequently transmitted one or the other pathogen to test plants (Catherall, 1970; Plumb & Medaiyedu, 1976). Infection is normally restricted to autumn-sown crops and is favoured by mild autumns. Both diseases can be lethal and on a few occasions infection with EWSM has been widespread, causing losses of 10-25%. More commonly a few isolated plants are infected, which may go unnoticed (Plumb, 1971b).

Two of the soil-borne, fungus-transmitted viruses prevent the economic cultivation of oats where they occur. Oat mosaic virus (OMV) was identified some time ago (Macfarlane *et al.*, 1968) but oat golden stripe virus (OGSV) has only recently been recognized (Plumb *et al.*, 1977). Although both viruses are thought to be transmitted by the plasmodiophoraceous fungus *Polymyxa graminis*, infection by OMV is usually more widespread than OGSV, which occurs only in small patches of plants also infected with OMV. Losses due to OMV can be considerable (Hayes & Lewis, 1975) and plants infected by OMV and OGSV are damaged more than plants infected by OMV alone (Plumb & Macfarlane, 1978). Infectivity is retained in soil for at least 10 years but the new range of fungicides, which are active against phycomycetes, including fungi that are virus vectors (Horak & Schlosser, 1978), may give some control, although it is unlikely that they will be sufficiently effective to eradicate fungal vectors in a field because of the difficulty of treating soil. If oats are to be grown, use of cultivars showing some resistance may be a more effective method of control (Catherall & Boulton, 1979).

   Barley yellow mosaic virus (BYMV) was first reported in Britain in 1980 (S. Hill, personal communication). While it is soil-borne and apparently fungus-transmitted, its distribution and the damage it causes are not yet known.

BARLEY YELLOW DWARF VIRUS

When compared with the viruses already described it is easy to see why so much interest has been shown in the control of barley yellow dwarf virus (BYDV). The aphid vectors of BYDV are ubiquitous and frequently present in large numbers. Most grasses are infected (Doodson, 1967; Plumb, 1977b) and act as important overwintering reservoirs. In consequence BYDV infection of cereals is often common. In 1967-70 and 1974-7, 19-89% of spring barley crops, and in 1971-77, 18-75% of winter wheat crops, were infected by BYDV (King, 1977). However, the factors that determine whether infection of cereals is sufficient to justify control are many and complex. All cereal cultivars commercially available are susceptible to BYDV and can be severely damaged if infected when young. Yield loss can be as much as 80-90% when infection is at growth stage (GS, Zadoks *et al.*, 1974) 10-12 but losses decline to insignificance if infection occurs when plants are "in boot" (GS 40-50) (Doodson & Saunders, 1970). Winter barley and winter oats are very susceptible and can be severely damaged; winter wheat and spring cereals are less damaged. Serological tests have shown that BYDV is a generic name covering many strains or variants (Aapola & Rochow, 1971; Clark *et al.*, 1979), some differing sufficiently to be considered distinct viruses. Strains also differ in efficiency of transmission by different aphid vectors, the damage they cause and in their geographical distribution in cereals (Plumb, 1977a).

Vector occurrence

Eighteen aphid vectors of BYDV are known (Plumb, 1974; A'Brook, 1978) but only three, *Rhopalosiphum padi, Macrosiphum (Sitobion) avenae* and

Metopolophium dirhodum, are of practical interest, although in some years *R. insertum* is important in Wales (A'Brook, 1978). The Rothamsted Insect Survey (RIS) showed that there are usually three migrations of cereal aphids each year (Taylor, 1977). The first is in late spring or early summer when aphids move from their winter hosts to cereals, the second corresponds to the major period of crop growth and to aphid dispersal associated with ripening in July and the third, in September and October, is of aphids, often sexual forms, seeking their primary hosts. The proportion of each species in these flights differs from season to season and site to site (Taylor, 1977) but *R. padi* and *M. (S.) avenae* are normally the most common species in the early migration, while *M. (S.) avenae* and *M. dirhodum* predominate in July, and *R. padi* and *R. insertum* in autumn.

Aphid infectivity

Aphids are the only means of spreading BYDV and tests of their infect-ivity over 11 years at Rothamsted showed that if they are divided into those caught before September (spring and summer migrations) and those caught later (autumn migration) the maximum percentage infective of any species was 11.6% of *R. padi* in autumn 1973 (Table 2). No *M. (S.) avenae* or *M. dirhodum* caught later than September have ever carried BYDV and in 5 out of 10 years neither did *R. padi*. In spring and summer, individuals of all three species were infective in most years

Table 2. Catches of cereal aphids in the Insect Survey trap at Rothamsted, and the proportion and number infective in September, October and November, 1970-79.

Year	*Rhopalosiphum padi*			*Macrosiphum (Sitobion) avenae**	*Metopolophium dirhodum**
	Number caught	% infective	Number infective	Number caught	Number caught
1970	2855	0	0	0	0
1971	1859	0.8	15	5	1
1972	842	6.2	52	12	3
1973	1393	11.6	161	21	4
1974	246	0	0	1	0
1975	47	0	0	0	0
1976	415	0	0	7	0
1977	2153	0	0	11	4
1978	2254	1.2	27	14	7
1979	1611	2.6	42	4	0

* No *M. (S.) avenae* or *M. dirhodum* were infective.

(Table 3). However, both population size and the percentage infective
differ between regions. The percentage of autumn migrants carrying
virus is generally greater at Aberystwyth (A'Brook, 1978) than at
Rothamsted; at Long Ashton in 1976, infective aphids were caught
earlier and a larger percentage were infective than at Rothamsted but
in 1977 the reverse was true (Smith et al., 1977, 1978). The date on
which infective aphids invade the crop also influences infection,
especially of spring-sown cereals. At Rothamsted in all years since
1969, except 1975 and 1978, the first infective aphid has been *R. padi*,
usually caught in May. The earliest infective aphid caught was *M. (S.)
avenae* on 13 May in 1975 (Plumb, 1976, 1977a).

Table 3. The proportion and number infective of cereal aphids in
each year up to the beginning of September, 1969-79.

| Year | Rhopalosiphum padi | | Macrosiphum (Sitobion) avenae | | Metopolophium dirhodum | | |
	%	Number*	%	Number*	%	Number*	Total
1969	9.5	55	5.9	125	2.7	10	190
1970	8.2	91	7.3	180	4.1	131	402
1971	1.4	24	3.5	112	2.6	43	179
1972	2.5	4	7.4	93	3.1	22	119
1973	11.1	38	3.6	23	0	0	61
1974	0	0	0.9	4	4.8	6	10
1975	11.5	117	3.0	46	5.1	19	182
1976	3.2	39	3.1	154	1.3	14	207
1977	4.9	225	4.7	163	3.8	21	409
1978	0	0	0.5	3	0.2	1	4
1979	0	0	0.8	9	0.1	15	24

* Derived by calculation from number caught in the Rothamsted Insect
Survey trap and % infective.

There is no apparent relationship between the number or percentage
of infective cereal aphids caught in the autumn of one year and in
the following summer, or between summer and autumn of the same year.
Although cold spring weather has been associated with large populations
of aphids in summer (Vickerman, 1977) this may bear little relation to
BYDV infection. For example, only 15 (0.1%) of the extraordinary
numbers of *M. dirhodum* caught at Rothamsted in 1979 transmitted BYDV
(Table 3). However, some generalizations can be made. In most years
infective aphids will not be detected at Rothamsted before the second
half of May and often none will be caught before the middle of July.
In many years no infective aphids will be caught in autumn; at

Aberystwyth no infective aphids were caught in autumn in 3 years out of 8 (A'Brook, 1978) and at Rothamsted none in 5 years out of 10 (Table 2).

BYDV IN AUTUMN-SOWN CROPS

Experiments were done at Rothamsted to investigate interactions between sowing date and BYDV on winter oats from 1972 to 1977, and on winter wheat from 1974 to 1977. In most years plots were sown in September, October and November and either treated with phorate granules at 5 kg a.i./ha at sowing or sprayed with menazon at 280 ml a.i./ha the following spring. In both 1972/3 and 1973/4 untreated oats yielded more, and were infected less, by BYDV when sown in November than when sown in September. Autumn and spring applied pesticides decreased BYDV incidence and increased yield on September-sown plots in 1972/3 by 0.31 t/ha (6.3%) and 0.35 t/ha (7.1%) respectively (Plumb, 1977a). The autumns of 1972 and 1973 were the last in which many aphids were infective (Table 2).

The greater risk of BYDV infection resulting from early autumn sowing was also demonstrated by the work of the Agricultural Development and Advisory Service (ADAS) aerial photography unit at Cambridge (Hooper, 1978). Crops sown later than mid-October showed little infection whereas earlier sown crops were often extensively infected by BYDV.

A premise of the experiments at Rothamsted was that September-sown crops must be protected as soon as they emerge if virus infection is to be controlled. Experiments done by ADAS in Pembrokeshire, while demonstrating that chemical control of aphids could be profitable, did not support this assumption. In one of two experiments sown in autumn 1972, phorate applied topically in December to oats sown in November increased yield by 0.23 t/ha (25%) but this was much less than the increase that resulted from spraying with dimethoate in December and March, which increased yield by 2.75 t/ha (305%). Similarly, dimethoate granules applied in December gave an increase of 0.4 t/ha (57%) and demeton-S-methyl sprays in December and March increased yield by 3.12 t/ha (446%). In 1973/4 dimethoate granules applied to oats shortly after emergence also increased yield (by 0.73 t/ha, 24%) but this was again less than the increase given by sprays of dimethoate in October and April (1.06 t/ha, 35%). A parallel experiment on winter barley suggested that of the two sprays only that in April was beneficial, as a single spray in October had no effect on yield whereas sprays in October and April increased yield by 1.23 t/ha (40%) (C. Guile, personal communication). As at Rothamsted, there were insufficient infective aphids in subsequent years to confirm these early observations. Even in 1977 when 90% of plants were infested with R. padi in October a spray had no effect on yield as few carried BYDV.

In a 1977 ADAS trial in Devon an oat crop sown on 15 September was

sprayed once with demeton-S-methyl (420 ml a.i./ha) on 18 November
when 97% of plants were infested with almost nine aphids/plant.
Sprayed plots, in which BYDV infection was minimal, outyielded
unsprayed, in which infection was widespread, by 1.38 t/ha (32%)
(M. Saynor, personal communication). The premise that early autumn-
sown crops need protecting as soon as they emerge no longer seems
tenable: what is not known is how long after sowing a spray can be
delayed. It seems likely that this will differ from year to year
and depend partly upon weather conditions in November and December
and partly on the success of aphids in surviving overwinter on the
crop and spreading virus in the spring. However, the denser, early-
sown crops may provide a more congenial environment overwinter than
crops sown later. One conclusion from these experiments might be
that to avoid infection it is best to sow late, but most farmers
are likely to prefer to accept the risk of BYDV rather than the
risk of a diminution in yield potential or not being able to sow at
all.

Another more general conclusion is that local variations are
important. For example, in autumn 1977 BYDV was widespread on early-
sown crops in Devon and East Anglia but not at Long Ashton or
Rothamsted. The largest and most prolonged autumn migration of cereal
aphids in the history of the RIS was in autumn 1978, but there were
many fewer aphids at Rothamsted than elsewhere and only 1.2% of the
autumn migration of *R. padi* was infective. Neither autumn nor spring
applied pesticides had any effect on virus infection or yield of
early-sown winter barley, although aldicarb applied at sowing
slightly increased the yield of September-sown winter wheat (0.25
t/ha, 2.6%). By contrast sprays applied in October or November to
early-sown wheat in East Anglia (D. Greaves, personal communication)
and barley in Devon (M. Saynor, personal communication) gave good
control of BYDV, with some treated crops outyielding untreated by
more than 80%.

If the percentage of *R. padi* infective is multiplied by the number
caught in the nearest RIS trap the result gives a measure of the
potential for BYDV infection. At Rothamsted in autumn 1978 this was
27 and did not apparently constitute a risk, whereas in autumn 1972
and 1973 figures of 52 and 161 (Table 2) seem to have justified
treatment.

BYDV IN SPRING-SOWN CROPS

When chemicals were used to control aphids and BYDV on spring barley
in 1971-1975, phorate granules at sowing or subsequent menazon sprays
occasionally increased yield (Table 4) and there was some correlation
between yield response, aphid numbers, percentage infective and the
date of capture of the first infective aphid in each year (Tables 3
and 4). In 1971, when infective aphids were fairly numerous and
migrated early, both granules and an early spray increased yield.
From 1972-1974 infective aphids were fewer and caught later. As a

result in 1972 and 1973 aphicides were of little benefit although in 1974 later sowing than usual seems to have resulted in a small beneficial effect of phorate granules. Although infective aphids occurred slightly earlier in 1975, and were as numerous as in 1971, chemicals had no effect, probably because the crop was sown early. Other, later-sown, crops at Rothamsted were much affected by virus.

Table 4. Effects on yield (t/ha) of chemical treatments applied to spring barley to control aphids and BYDV.

| | Sowing date | | | | |
| | 1971 | 1972 | 1973 | 1974 | 1975 |
	12 March	29 March	17 March	2 April	27 February
First infective aphid	19 May	22 May	30 May	4 July	15 May
Untreated	5.72	6.26	6.34	6.06	4.68
Phorate granules 5 kg a.i./ha (at sowing)	6.18	6.54	6.43	6.26	3.92
Menazon spray 280 ml a.i./ha early (date)	5.95 (24/5)	6.56 (1/6)	6.12 (8/6)	6.11 (20/6)	4.14 (13/6)
Menazon spray late (date)	5.77 (23/6)	6.46 (30/6)	6.17 (9/7)	6.20 (23/7)	4.48 (9/7)
SED	0.077	0.155	0.124	0.105	0.255

From 1976, experiments also tested the effects of sowing date (Plumb, 1977a), and from them and the results from previous years several general conclusions can be drawn.

1. Late-sown spring crops (mid-April onwards) are generally twice as likely to be infected by BYDV and will carry twice as many aphids as those sown earlier.
2. Pesticides give proportionately greater benefits on late-sown crops.
3. At Rothamsted widespread infection with BYDV is very unlikely, even on crops sown late.
4. Aphids can be directly damaging to barley and large increases in yield can result from their chemical control.

It is not known why aphids should be more numerous on late-sown barley crops but it may result from aphids either being preferentially attracted to them, as has been reported for other crops (A'Brook, 1968), or multiplying more rapidly on young than on older plants. Because BYDV can be very damaging if infection occurs at early growth stages a granular treatment applied at sowing to barley sown in late April in Eastern Britain may be beneficial. In the west, where sowing is generally later (Plumb, 1977a) and where the severer isolates of BYDV are most common (Plumb, 1971a), chemical treatment might seem to be more frequently justified. However, in the few experiments that have tested the effects of pesticides on spring-sown crops in western areas, no significant increases in yield were obtained at Seale Hayne (Devon) in 1972-4 or at Rosemaund (Hereford) in 1972-3. This may be because the aphicides did not effectively protect the crop during the slightly larger and more prolonged spring aphid migration in the west (Plumb, 1977a).

CONCLUSIONS

Clearly chemicals have a role to play in controlling BYDV and possibly will have for the soil-borne, fungus-transmitted OMV, OGSV and BYMV in the future.

The risk of BYDV infection of autumn-sown crops is great for those sown in September and negligible for those sown after the middle of October. While granular pesticides applied at sowing can restrict BYDV infection, their use does not seem justified as the likelihood of infection can be assessed after sowing, by monitoring aphid numbers and the percentage that are infective. If infection seems likely then a spray with a moderately persistent aphicide in early November should give good control but if conditions prevent autumn spraying, a treatment the following spring may be beneficial, especially in areas with mild winters.

While sowing date is more likely to be determined by weather and organizational considerations than the threat of BYDV, the need to use pesticides may be minimized by sowing those fields with a past history of BYDV later rather than earlier.

In Eastern Britain, spring-sown cereals are rarely extensively infected with BYDV even when sown late, so the use of a granular pesticide is rarely justified, except perhaps when crops are sown after the middle of April following a mild winter. In the west of Britain the little evidence available suggests that chemicals are less effective than in the east.

REFERENCES

Aapola A.I.E. & Rochow W.F. (1971) Relationships among three isolates of barley yellow dwarf virus. *Virology* 46, 127-41.
A'Brook J. (1968) The effect of plant spacing on the numbers of aphids trapped over the groundnut crop. *Annals of Applied*

Biology 61, 289-94.

A'Brook J. (1978) Infectivity of barley yellow dwarf virus vectors. *Report of the Welsh Plant Breeding Station for 1977*, pp. 181-5.

Benigno D.A. & A'Brook J. (1972) Infection of cereals by cocksfoot mottle and phleum mottle viruses. *Annals of Applied Biology* 72, 43-52.

Catherall P.L. (1970) Oat sterile dwarf virus. *Plant Pathology* 19, 75-8.

Catherall P.L. & Boulton R.E. (1979) Reaction of some winter oat cultivars to oat mosaic and oat tubular viruses. *Plant Pathology* 28, 57-60.

Clark M.F., Bates D., Plumb R.T. & Lennon E. (1979) Virus detection and characterization: Barley yellow dwarf virus. *Report of the East Malling Research Station for 1978*, p. 101.

Doodson J.K. (1967) A survey of barley yellow dwarf virus in S.24 perennial ryegrass in England and Wales, 1966. *Plant Pathology* 16, 42-5.

Doodson J.K. & Saunders P.J.W. (1970) Some effects of barley yellow dwarf virus on spring and winter cereals in field trials. *Annals of Applied Biology* 66, 361-74.

Hayes J.D. & Lewis D.A. (1975) Oat Studies a) Soil borne winter oat mosaic virus (OMV). *Report of the Welsh Plant Breeding Station for 1974*, p. 31.

Hooper A.J. (1978) Aerial Photography. *Journal of the Royal Agricultural Society of England* 139, 115-23.

Horak I. & Schlösser E. (1978) Wirkung von prothiocarb auf *Polymyxa betae* und *Olpidium brassicae*. *Mededeelingen van der Faculteit Landbouwetenschappen Rijksuniversiteit Gent* 43, 979-87.

Kassanis B. & White R.F. (1977) Possible control of plant viruses by polyacrylic acid. *Proceedings 1977 British Crop Protection Conference - Pests and Diseases* 3, 801-6.

King J.E. (1977) The incidence and economic significance of diseases in cereals in England and Wales. *Proceedings 1977 British Crop Protection Conference - Pests and Diseases* 3, 677-87.

Lewis T. (1977) Prospects for monitoring insects using behaviour controlling chemicals. *Proceedings 1977 British Crop Protection Conference - Pests and Diseases* 3, 847-56.

Macfarlane I., Jenkins J.E.E. & Melville S.C. (1968) A soil-borne virus of winter oats. *Plant Pathology* 17, 167-70.

Mulligan T.E. (1959) Ryegrass mosaic. *Report of the Rothamsted Experimental Station for 1958*, p. 100.

Plumb R.T. (1971a) The control of insect-transmitted viruses of cereals. *Proceedings of the 6th British Insecticides and Fungicides Conference* 1, 307-13.

Plumb R.T. (1971b) European wheat striate mosaic disease in 1970. *Plant Pathology* 20, 120-2.

Plumb R.T. (1974) Properties and isolates of barley yellow dwarf virus. *Annals of Applied Biology* 77, 87-91.

Plumb R.T. (1976) Barley yellow dwarf virus in aphids caught in suction traps, 1969-73. *Annals of Applied Biology* 83, 53-9.

Plumb R.T. (1977a) Aphids and virus control on cereals. *Proceedings 1977 British Crop Protection Conference - Pests and Diseases* 3,

903-13.

Plumb R.T. (1977b) Grass as a reservoir of cereal viruses. *Annales de Phytopathologie* 9, 361-4.

Plumb R.T., Catherall P.L., Chamberlain J.A. & Macfarlane I. (1977) A new virus of oats in England and Wales. *Annales de Phytopathologie* 9, 365-70.

Plumb R.T. & Lennon E. (1980) Agropyron mosaic virus in wheat. *Report of the Rothamsted Experimental Station for 1979*, Part 1, p. 173.

Plumb R.T. & Macfarlane I. (1978) Cereal Diseases. *Report of the Rothamsted Experimental Station for 1977*, Part 1, pp. 212-3.

Plumb R.T. & Medaiyedu J.A.O. (1976) Diseases of grass and forage crops. *Report of the Rothamsted Experimental Station for 1975*, Part 1, pp. 259-60.

Slykhuis J.T. (1976) Virus and virus-like diseases of cereal crops. *Annual Review of Phytopathology* 14, 189-210.

Smith B.D., Kendall D.A., Singer M.C., Halfacree S. & Mathias L. (1977) Cereal aphids and spread of barley yellow dwarf virus. *Report of the Long Ashton Research Station for 1976*, p. 97.

Smith B.D., Kendall D.A., Hazell S., Mathias L., Hammock P. & March C. (1978) Cereal aphids and spread of barley yellow dwarf virus. *Report of the Long Ashton Research Station for 1977*, pp. 99-100.

Taylor L.R. (1977) Aphid forecasting and the Rothamsted Insect Survey. *Journal of the Royal Agricultural Society of England* 138, 75-97.

Vickerman G.P. (1977) Monitoring and forecasting insect pests of cereals. *Proceedings 1977 British Crop Protection Conference - Pests and Diseases* 1, 227-34.

Zadoks J.C., Chang T.T. & Konzak C.F. (1974) A decimal code for the growth stages of cereals. *Weed Research* 14, 415-21.

Section 3
Husbandry

Chairman's comments

E. LESTER
Rothamsted Experimental Station,
Harpenden, Hertfordshire

The title of this section is extremely broad and clearly most, if not all, of the preceding chapters could just as appropriately have been included here. Inevitably the authors of the chapters which follow have written from their own experience of particular soils, locations and climates and it is worth reminding the reader that extrapolation beyond these confines should usually be restricted to principles.

Cereal production, especially in north-west Europe, has become much more intensive in the last 20 years so that it is appropriate that the first chapter in this section deals with rotations. Although intensification of production may have been a consequence of economic pressures it has largely been made possible by the development of chemical weed-killers which have also allowed growers to adopt reduced cultivation methods. The rotation and cultivation methods are prime determinants of the threat posed by soil-borne and trash-borne pathogens.

The concept of biological control is relatively recent but its principles have been exploited for centuries, perhaps unconsciously, in the use of rotations to control soil-borne diseases. Even now the biological mechanisms involved are only partly understood but they are slowly yielding to painstaking investigations. The development of chemicals for the control of cereal root diseases is still very much in its infancy. By contrast, much progress has been made with the development of chemicals which control leaf pathogens and minimize the losses they cause. The availability of these compounds also allows growers to adopt some practices which the risk of diseases formerly made unacceptable, for example the early sowing of winter cereals. However, there is the risk that there may be increased damage from pathogens for which adequate control measures are still not available. Consequently inte-

grated control of individual diseases is an inadequate objective that
may involve contradictory decisions. Integrated protection of the *crop*
is a more desirable aim, recognizing that freedom from disease is
rarely necessary even if it is possible. However, such an objective
will be achieved only when each component of the disease (and pest)
spectrum is better understood. The following chapters provide some
measure of our present knowledge and point the way to its improvement.

Cropping systems in relation to soil-borne and trash-borne diseases of cereals

R. D. PREW

Rothamsted Experimental Station,
Harpenden, Hertfordshire

INTRODUCTION

Numerous factors influence a farmer's choice of rotation, including climate, diseases, labour, machinery, personal preferences, pests, prices, soil-type and weeds. The relative importance of these factors will differ greatly between farms and farmers. Therefore it is unrealistic to design cropping strategies solely to minimize diseases; rather, it is more appropriate to consider the part diseases play, among all the other contributory factors, in determining the choice of cropping sequence.

In recent years economic pressures, especially the large capital costs of modern agriculture, have forced many farmers to simplify their farming systems and to specialize in fewer enterprises. These pressures, together with technological advances, particularly in the field of chemical control, have resulted in a large increase in the number of farms in England and Wales classified as "mostly cereals"; the proportion of the total national area of wheat and barley grown on these farms has doubled in 10 years (Table 1). Unfortunately the censuses on which these figures are based provide no information on the proportion of wheat and barley crops grown successively, although data from the 1974 survey of fertilizer use in England and Wales (M.G. Hills, personal communication) showed that 29% of wheat followed one previous cereal and 21% followed two or more cereals. Increasingly frequent cropping with cereals increases the risk of damage by some diseases. This chapter considers the problems that the main soil-borne and trash-borne diseases, eyespot (*Pseudocercosporella herpotrichoides*), take-all (*Gaeumannomyces graminis* var. *tritici*), septoria (*Septoria nodorum*) and rhynchosporium (*Rhynchosporium secalis*), pose to a farmer wishing to increase the proportion of cereals in his rotation to more than 50%.

CEREAL CROPPING SEQUENCES

If the starting point in the rotation is taken as the first wheat crop,

Table 1. Farms classified as "mostly cereals" in England and Wales (Anon., 1967, 1976).

	1964	1974
Total number of farms	157,391	129,738
Farms classified as "mostly cereals"	6,345	9,053
% of total wheat area on "mostly cereal" farms	17	28
% of total barley area on "mostly cereal" farms	11	29

then the farmer has to consider whether the length or type of the preceding break crop will have an effect on diseases in this first wheat. For eyespot the length of break rather than the crop used is the important factor. The eyespot fungus survives on infected debris for up to 3 years (Macer, 1961) and Glynne & Moore (1949) showed that a break of more than 1 year was needed to ensure good control of this disease. By contrast, a 1-year break with a range of different crops can give satisfactory control of take-all (Prew & Dyke, 1979) (Table 2). Oats, unlike barley, is not a host crop for take-all except where the oat strain of the take-all fungus (*Gaeumannomyces graminis* var.

Table 2. The effects of previous cropping and rates of fertilizer nitrogen on take-all of winter wheat.

Previous crop	Take-all rating* Nitrogen (kg/ha)			
	0	50	100	150
Barley	168	154	112	120
Barley undersown with trefoil	170	166	142	114
Oats	26	23	7	3
Clover	24	13	17	11
Beans	15	16	12	7
Maize	20	8	5	3

* Take-all rating = % plants with slight take-all + 2 (% plants with moderate take-all) + 3 (% plants with severe take-all).

avenae) is present. Grass leys can be an effective break provided

rhizomatous grass weeds are not present (Wehrle & Ogilvie, 1955).
However, a grass break gives poorer control of septoria than do other
break crops (King, 1977), although as this disease is also frequently
seed-borne (Machacek, 1945; Hewett, 1975; Jenkyn & King, 1977) no
break can ensure freedom from septoria.

Deciding what crop should follow this first wheat crop can be
difficult. If a second wheat crop is to be grown eyespot, take-all
and septoria may all be damaging. Whilst eyespot and septoria can be
controlled by using fungicides and resistant cultivars, this is not
possible for take-all. Increase in take-all is so variable that prediction of damage to a second wheat crop is very difficult (Hornby,
1978). Experience has shown that some soil types are less prone to
severe take-all than others. However, in 1979 second wheats were
damaged even on soils thought safe for them. Despite these results,
a second wheat is an acceptable risk on some soils. There is evidence
that if the two wheats follow a grass break then, because of the
build-up of another fungus (*Phialophora radicicola* var. *graminicola*)
on the grass roots, the chances of avoiding damaging take-all are
improved (Deacon, 1973).

In an experiment at Rothamsted (Prew, 1977) first wheats after
different breaks all had very little take-all, but there were considerable differences in the incidence of take-all in the second wheats,
with severe infection after fallow or lucerne but not after a grass/
clover ley (Table 3). The second wheat after grass/clover yielded
more than twice as much grain as the second wheat after fallow. The
effect of the grass/clover was only to delay the development of take-
all and third wheats after lucerne and after grass/clover were
equally infected. A bioassay technique (Slope *et al.*, 1979), used to
measure the changes in population of *G.g. tritici* and *P.r. graminicola* in the soils of this experiment, showed that, despite the small
incidence of take-all in the first wheats, the soil population of
G.g. tritici had increased by the end of that season and there were
then differences between treatments similar to those observed on the
roots of the second wheat crops (Table 4).

Table 3. The incidence of take-all in first, second and third
wheat crops (W_1, W_2 and W_3) after different crops.

Previous crop	Take-all rating*		
	W_1	W_2	W_3
Grass/clover	4	38	115
Lucerne	3	149	128
Fallow	10	240	84

* See Table 2.

Table 4. The incidence of *Gaeumannomyces graminis* var. *tritici* and *Phialophora radicicola* var. *graminicola* in soils after different crops (W_1, W_2 and W_3 represent successive wheat crops following the break crops shown).

Previous crop	% Soil cores with *G.g. tritici*			*P.r. graminicola* rating*		
	W_1	W_2	W_3	W_1	W_2	W_3
Grass/clover	8	31	61	188	125	78
Lucerne	46	68	82	72	108	55
Fallow	61	61	81	38	55	52

* Maximum rating 300, details in Slope *et al.* (1979).

The negative relationship between populations of *G.g. tritici* and *P.r. graminicola* in the first wheat soils is also shown in Table 4. The details of this relationship and its mechanisms remain obscure as shown by Slope *et al.* (1979) who also reported that large populations of *P.r. graminicola* are present in soils after crops other than grass leys. It is possible that some control of take-all may be achieved by introducing or maintaining large populations of *P.r. graminicola* in soil by inoculation or by use of grass catch crops.

The next stage of the rotation poses even greater problems, for if a third wheat is grown it will on many, if not most, soils be at considerable risk from take-all. Although the familiar pattern of take-all build-up and decline (Slope *et al.*, 1970) seems to occur generally, the peak may vary in both severity and timing. It is therefore wise to grow spring barley as the third cereal, because it will almost certainly be less infected with take-all than a winter wheat. Furthermore, the ameliorating effect of applying nitrogen fertilizers is believed to be greater on spring barley than on winter wheat. Although there is little direct experimental evidence to support these statements, Cunningham (1965) showed that spring-sown barley is less infected than spring wheat. In experiments at Rothamsted (Prew & Dyke, 1979), spring barleys grown as second or fourth cereals, in the same year, yielded better and had less take-all than wheat grown as a third cereal the previous year (Table 5). Proportionately, take-all on the barley was decreased more by increasing use of nitrogen.

At this point in the rotation many farmers return to a break crop. However, if a longer run of cereals is to be grown, continuing in spring barley during the peak take-all period seems best, although rhynchosporium may then become a problem. This disease can be seed-borne (Caldwell, 1937) but infected debris on the soil surface is

Table 5. Grain yield (t/ha) and take-all of spring barley and
winter wheat (means of two experiments).

| | | Nitrogen (kg/ha) | | | |
		0	50	100	150
Spring barley	Yield	3.71	5.60	5.76	5.71
2nd cereal	Take-all rating*	94	57	33	28
4th cereal	Yield	3.44	5.34	5.96	5.71
	Take-all rating	90	62	54	42
Winter wheat	Yield	1.78	2.71	3.54	3.24
3rd cereal	Take-all rating	163	142	108	127

* See Table 2.

probably the main source of inoculum (Evans, 1969; Kay & Owen, 1973).
However, even surveys done in different regions in the same year show
contrasting effects of previous cropping on the incidence of this
disease (James, 1969; Evans, 1969; Melville & Lanham, 1972). As little
inoculum is needed to start an epidemic (Jenkins & Jemmett, 1967) when
environmental conditions are favourable (Polley, 1971), the frequency
of suitable conditions for infection seems to be at least as important
as the frequency of past cereal cropping. Resistant cultivars and
fungicides should give adequate control even in areas particularly
suitable for this pathogen.

With the recent increase in area sown to winter barley, there may be
a temptation to substitute winter barley for spring barley during the
period of greatest risk from take-all. While there is no experimental
work directly comparing the incidence of diseases on them there may be
dangers in such a substitution. With successive winter barley crops
the period for the fungi to survive between crops is shorter than with
spring barley and the period during which the crop is available for
infection is longer, so that the incidence of eyespot, take-all and
rhynchosporium can be expected to be greater. Certainly winter barley
can be severely infected with both eyespot and take-all (Slope & Cox,
1966). Scott *et al.* (1975) reported winter barley to be less susceptible to eyespot than winter wheat, but the wheat cultivar they used
was Hybrid 46, which is very susceptible. Most recently introduced
wheat cultivars are more resistant and have stronger straws so that
the weak-strawed winter barley cultivars may be more at risk from
eyespot than winter wheat. Until experimental evidence is available
it seems sensible not to use winter barley as a substitute for spring
barley at this point in the crop sequence.

Once take-all decline is established, winter wheat can again be grown. However, it is difficult for the farmer to know when this stage is reached. To do so with any confidence it is necessary to assess the amount of take-all on the roots of crops at all stages in the rotation. Even so, identifying the decline phase can be difficult because amounts of take-all show large seasonal variations. If successive wheat crops were then grown, they would be at risk from septoria and eyespot. Alternating wheat and barley crops can decrease the incidence of take-all and eyespot (Cunningham, 1975, 1980). However, in long runs of winter cereals, sprays for the control of eyespot are almost certain to be needed whichever crops are grown. Use of carbendazim-generating fungicides to control eyespot can lead to an increase in sharp eyespot (*Rhizoctonia solani*) (Jenkyn & Prew, 1973; Prew & McIntosh, 1975) and it is possible that frequent use of these fungicides in long runs of cereals could lead to a cumulative increase in the incidence of sharp eyespot.

CONCLUSIONS

The diseases discussed in this chapter may influence the husbandry of the crops grown but with the present availability of fungicides and resistant cultivars, only take-all is likely to be an important factor in determining the choice of cropping sequence. A farmer wishing to grow mainly cereals and minimise the effects of take-all has two options:

A.   WW* WW   SB   Br

B. (i) On half of the arable acreage
         WW   Br   WW   Br

    (ii) On the remainder of the arable acreage
         WW   WW   SB   SB   SB   WW   WW   WW   WW   Br

         *WW, Winter wheat; SB, Spring barley; Br, Break crop

In the first option (A) both the second wheat and the spring barley are at some risk from take-all. The sequence gives 75% cereals with 25% first wheat. The combination of rotations in option (B), offers both (i) a traditional rotation in which high input, high yielding first wheats are grown and (ii) a long run of cereals on the rest of the arable area, which takes maximum advantage of take-all decline. These rotations together give 70% cereals with 30% first wheats. However, many factors play a part in the choice of rotation so although the options described above may limit the effects of take-all they may not be the most suitable when all other factors are considered.

REFERENCES

Anon. (1967) Ministry of Agriculture, Fisheries and Food. Farm

Classification in England and Wales for 1964-65. London, H.M.S.O.

Anon. (1976) Ministry of Agriculture, Fisheries and Food. Farm Classification in England and Wales for 1974. London, H.M.S.O.

Caldwell R.M. (1937) Rhynchosporium scald of barley, rye and other grasses. *Journal of Agricultural Research* 55, 175-98.

Cunningham P.C. (1965) Investigation of the reaction of wheat and barley to *Ophiobolus graminis*. *Research Report for 1964, Plant Sciences and Crop Husbandry*, An Foras Taluntais, pp. 97-8.

Cunningham P.C. (1975) Some consequences of cereal monoculture on *Gaeumannomyces graminis* (Sacc) Arx & Oliver and the take-all disease. *EPPO Bulletin* 5, 297-317.

Cunningham P.C. (1980) Differential host preference in *Pseudocercosporella herpotrichoides* with spring sown wheat and barley monoculture. *Annals of Applied Biology* 94, 33-9.

Deacon J.W. (1973) Control of the take-all fungus by grass leys in intensive cereal cropping. *Plant Pathology* 22, 88-94.

Evans S.G. (1969) Observations on the development of leaf blotch and net blotch of barley from barley debris, 1968. *Plant Pathology* 18, 116-8.

Glynne M.D. & Moore F.J. (1949) Effect of previous crops on the incidence of eyespot in winter wheat. *Annals of Applied Biology* 36, 341-51.

Hewett P.D. (1975) *Septoria nodorum* on seedlings and stubble of winter wheat. *Transactions of the British Mycological Society* 65, 7-18.

Hornby D. (1978) The problems of trying to forecast take-all. *Plant Disease Epidemiology* (Ed. by P.R. Scott & A. Bainbridge), pp. 151-8. Blackwell Scientific Publications, Oxford.

James W.C. (1969) A survey of foliar diseases of spring barley in England and Wales in 1967. *Annals of Applied Biology* 63, 253-63.

Jenkins J.E.E. & Jemmett J.L. (1967) Barley leaf blotch. *Quarterly Review of the National Agricultural Advisory Service* No. 75, 127-32.

Jenkyn J.F. & King J.E. (1977) Observations on the origins of *Septoria nodorum* infection of winter wheat. *Plant Pathology* 26, 153-60.

Jenkyn J.F. & Prew R.D. (1973) Activity of six fungicides against cereal foliage and root diseases. *Annals of Applied Biology* 75, 241-52.

Kay J.G. & Owen H. (1973) Transmission of *Rhynchosporium secalis* on barley grain. *Transactions of the British Mycological Society* 60, 405-11.

King J.E. (1977) Surveys of diseases of winter wheat in England and Wales 1970-75. *Plant Pathology* 26, 8-20.

Macer R.C.F. (1961) The survival of *Cercosporella herpotrichoides* Fron in wheat straw. *Annals of Applied Biology* 49, 165-72.

Machacek J.E. (1945) The prevalence of Septoria on cereal seed in Canada. *Phytopathology* 35, 51-3.

Melville S.C. & Lanham C.A. (1972) A survey of leaf diseases of spring barley in South-West England. *Plant Pathology* 21, 59-66.

Polley R.W. (1971) Barley leaf blotch epidemics in relation to weather conditions with observations on the overwintering of the disease on barley debris. *Plant Pathology* 20, 184-90.

Prew R.D. (1977) Studies on the spread, survival and control of take-all and other foot and root diseases of wheat and barley. Ph.D. Thesis, University of London, 157 pp.

Prew R.D. & Dyke G.V. (1979) Experiments comparing 'break crops' as a preparation for winter wheat followed by spring barley. *Journal of Agricultural Science, Cambridge* 92, 189-201.

Prew R.D. & McIntosh A.H. (1975) Effects of benomyl and other fungicides on take-all, eyespot and sharp eyespot diseases of winter wheat. *Plant Pathology* 24, 67-71.

Scott P.R., Hollins T.W. & Muir P. (1975) Pathogenicity of *Cercosporella herpotrichoides* to wheat, barley, oats and rye. *Transactions of the British Mycological Society* 65, 529-38.

Slope D.B. & Cox J. (1966) *Report of the Rothamsted Experimental Station for 1965*, p.124.

Slope D.B., Etheridge J. & Henden D.R. (1970) *Report of the Rothamsted Experimental Station for 1969*, Part 1, p.158.

Slope D.B., Prew R.D., Gutteridge R.J. & Etheridge J. (1979) Take-all, *Gaeumannomyces graminis* var. *tritici* and the yield of wheat grown after ley and arable rotations in relation to the occurrence of *Phialophora radicicola* var. *graminicola*. *Journal of Agricultural Science, Cambridge* 93, 377-89.

Wehrle V.M. & Ogilvie L. (1955) Effect of ley grasses on the carry over of take-all. *Plant Pathology* 4, 111-3.

Effects of cultivation methods on disease

D. J. YARHAM* & J. NORTON†
*Agricultural Development and Advisory Service,
Brooklands Avenue, Cambridge
†Department of Biology, Liverpool Polytechnic,
Byron Street, Liverpool

INTRODUCTION

For more than a 1000 years the mould-board plough has been the foundation of arable agriculture, not only facilitating the production of a
seed bed but also efficiently controlling weeds by burying them.
Unfortunately ploughing, although very effective, is also very costly
in both time and energy and, as labour and fuel costs have risen,
farmers have sought alternative methods of preparing land for cropping.
It was not, however, until effective herbicides became available that
such techniques began to offer a serious alternative to ploughing.
Stubble sprayed with paraquat or glyphosate, for example, can be
speedily cultivated with a tined implement, or seed can be sown directly
into a sprayed sward or stubble. Not surprisingly the area of cereals
grown without using a plough increases year by year.

There are four basic methods of preparing land for sowing:

1. Mould-board ploughing (MBP) in which the soil is inverted to a
   depth of about 20 cm and then further cultivated to produce a
   fine seed bed.
2. Deep cultivation (DC) in which a rigid-tined implement disturbs
   the soil to a depth of about 20 cm, again necessitating
   further cultivation.
3. Reduced cultivation (RC) where a relatively shallow-tined
   implement (10 cm) disturbs the soil just sufficiently to produce
   a seed bed.
4. Direct drilling (DD) in which the seed is sown directly into
   undisturbed soil. Specially constructed drilling equipment is
   normally used but if soil conditions are good a standard drill
   may be suitable.

LEAF DISEASES

By burying infected volunteer plants, ploughing decreases inoculum of
mildew *Erysiphe graminis* and rusts which overwinter on living tissue.

The origin of early attacks of yellow rust *Puccinia striiformis* can often be traced to volunteers in stubbles that remained unploughed when nearby autumn-sown crops had emerged (though nowadays a herbicide spray can cope with this nuisance as effectively as the plough).

Herbicides do not, however, decrease the inoculum of pathogens which survive on plant debris on the soil surface. Paraquat, for example, has no effect on the production of *Rhynchosporium secalis* spores on barley stubble (Stedman, 1977), and neither paraquat nor glyphosate affect the production of pycniospores of *Septoria nodorum* on wheat straw (Harris, 1979). Both diseases can, therefore, be aggravated by non-plough techniques. In five cultivation trials done by the Eastern Region of the Agricultural Development and Advisory Service (ADAS) in 1979 (all following a 1978 wheat crop) septoria was assessed on the second and third youngest leaves at growth stage (GS: Zadoks *et al.*, 1974)51-75. On plots that had been ploughed, given a reduced cultivation treatment, or direct drilled, average leaf areas affected were 17, 22 and 26% respectively. Prew (cited by Yarham & Hirst, 1975) obtained similar results with rhynchosporium on winter barley. Direct-drilled plots had seven times as much leaf blotch as ploughed plots in December and three times as much in February, though by May these differences had disappeared. Inoculum of trash-borne leaf diseases can, however, be decreased by straw burning. At Rothamsted direct-drilled winter barley assessed in January, February and April had respectively four, three and two times as much rhynchosporium where straw was not burnt as where it was (R.D. Prew, personal communication)

Increases in the severity of trash-borne leaf diseases where reduced cultivations are used (and where wheat follows wheat, or barley follows barley) have been demonstrated in many trials. Exceptions do occur, however, as in an ADAS trial on winter barley following winter barley done in Essex in 1979. On plots which compared mould-board ploughing, deep cultivations, reduced cultivations and direct drilling, amounts of debris on the soil surface during winter averaged 4, 10, 20 and 45 g/m^2 respectively, but in the following summer leaf areas affected by rhynchosporium on the four treatments averaged 34, 18, 21 and 16% respectively. Clearly, disease severity is not determined only by the amount of inoculum present and cultivar susceptibility. Another important factor is microclimate within the crop, which is influenced by plant vigour and crop uniformity. In the above experiment direct-drilled plots gave the smallest yields and were always the least vigorous. Direct-drilled winter barley crops often grow less vigorously than crops sown on ploughed land, an effect which may sometimes be aggravated by decreased mineralization of soil nitrogen in undisturbed soil. (Davies *et al.*, 1979).

The effects of cultivation treatment on crop vigour and microclimate are even more important with air-borne diseases, such as the rusts and mildews, than with splash-borne diseases. Rust and mildew spores are produced so prolifically and dispersed so readily that amounts of surviving inoculum are not critical. In trials, therefore, these diseases tend to be most severe in treatments with the most vigorous

and, therefore, most susceptible plants. In an experiment at Boxworth
Experimental Husbandry Farm (EHF) in 1979 there was much more mildew
on ploughed than on unploughed land (% area of 2nd youngest leaf on
16 July: MBP 21%, RC 8%, DD 4%). Similar effects were noted with
brown rust in a trial at Terrington EHF. At both sites disease severity
was inversely correlated with the numbers of fertile shoots per unit
area.

EYESPOT

The eyespot fungus (*Pseudocercosporella herpotrichoides*) survives, in
the absence of the growing crop, on dead stem bases. Stem bases
buried one year may be ploughed up the next year (or even the year
after) and still produce viable spores. As a result, where cereals are
grown intensively, ploughing will not decrease eyespot inoculum as
effectively as it will the inoculum of leaf diseases, even though there
will obviously be less debris on the surface of ploughed than direct-
drilled land. Although a very small amount of infected debris is
sufficient to initiate infection it has, nevertheless, been shown that
there is a relationship between severity of eyespot and the amount of
inoculum present in the winter (Cox & Cock, 1962). More eyespot might
therefore be expected in crops in non-ploughed than in ploughed fields.
In practice, however, there is no simple relationship between cultivation treatment and disease incidence. Numerous ADAS trials have shown
that effects of cultivation treatments on eyespot vary both between
years and sites. It is not uncommon, however, for eyespot to be less
severe in direct-drilled than in ploughed (and conventionally drilled)
plots, as results from a trial in Essex illustrate (Table 1). Brooks
& Dawson (1968) first noticed that there was often less eyespot in
direct-drilled crops, and suggested that this was because slow crop

Table 1. Effects of cultivation methods on incidence of eyespot at
Rawreth, Essex.

| | % plants with eyespot (spring) | | | % shoots with eyespot (summer) | | |
	MBP	RC	DD	MBP	RC	DD
1974				13	14	5
1975	85	82	48	57	13	3
1976	55	53	31	77	81	68
1977	57	45	39	82	80	50
1978	14	11	13	32	18	20
1979	15	7	6	35	39	34
Mean	45	40	27	49	41	30

establishment resulted in an unsuitable microclimate for disease development. However, in dry autumns, when direct-drilled crops often establish more rapidly than those sown after ploughing, they still often have less eyespot.

Alternatively, this effect of direct drilling may result from the change in plant growth habit. In an experiment which compared seed rates and precision versus conventional drilling, there was no effect of seed rate on infection with eyespot in conventionally-drilled plots in April, although by June there was more eyespot at higher seed rates (Table 2). On both dates there was less eyespot in precision drilled plots than in conventionally-drilled plots with the same seed rate. This effect was not related to initial plant populations or tiller numbers in spring. Throughout the winter and early spring, however, precision-drilled plants were much more prostrate than conventionally drilled plants (P. Gladders, personal communication). Direct-drilled plants, like precision-drilled plants, are often very prostrate because they are normally shallowly drilled. Even when they are sown more deeply, they tend to produce their crowns nearer to the soil surface than do plants in cultivated soil. However, the relationship between sowing depth, crown depth, plant habit and cultivation treatment is not absolutely clear (Table 3). The interaction between cultivation treatment, growth habit, and eyespot incidence was further investigated by taking plants in April and May 1979 from four cultivation trials and relating amounts of infection on individual plants to growth habit. The results confirm that there is an association between plant habit and eyespot incidence although this factor explains only a small proportion of the total variation in eyespot (Table 4).

Table 2.   Effects of seed rates and drilling methods on eyespot

Seed (kg/ha)	Drilling method	Number of plants/m^2 (22 Nov)	Number of shoots/m^2 (24 March)	Eyespot % plants (4 April)	%shoots (14 June)
110	Conventional	158	853	29	3.3
136	Conventional	194	1228	29	7.2
136	Precision	249	1192	9	3.3
152	Conventional	217	1087	31	9.2

Clearly the effects of cultivation on eyespot are complex. Ploughing usually decreases inoculum at the soil surface but only very occasionally is this sufficient to decrease the incidence of eyespot compared with that in otherwise comparable crops grown in unploughed fields. If sufficient inoculum remains on the soil surface to initiate an epidemic, disease incidence is more likely to reflect differences in crop growth. Differences in crop establishment and vigour may sometimes influence disease but current evidence suggests that plant habit also plays an important role.

Table 3. Crown depth, sowing depth and plant habit (mean of 20 plants/plot) (winter wheat – Margaretting, Essex, 1979).

Crown depth (mm)	Sowing depth (mm)	Angle between the two outermost tillers (2 April)	Cultivation treatment
7	22	42	DD
8	16	49	DD
9	18	45	RC
9	15	54	DD
11	17	31	DC
13	31	20	DC
25	38	13	DC
26	32	8	RC

Table 4. Relationships between eyespot infection and plant habit

Cultivar	Site	Number of plants	Equation (e =)	% variance accounted for	P
Winter wheat					
Sportsman	Rawreth	450	67–1.16a	25	0.001
Hobbit	Hundon	450	65–0.44a	5	0.02
Winter barley					
Maris Otter	Weeley	600	1–0.56a	31	0.001
Maris Otter	Hindringham	450	1–0.44a	19	0.001

(e, % shoots with eyespot ; a, angle between the two outermost tillers)

TAKE-ALL

Cultivation technique might be expected to affect incidence of root diseases by altering the distribution of inoculum. Fungal propagules normally dispersed down the soil profile by mould-board ploughing may, on unploughed land, remain in the surface layers where the seed is sown. Redistribution of inoculum may explain why, in one trial, there was less take-all in land ploughed to a depth of 20 cm than in land ploughed to 10 cm or land where ploughing to 30 cm had failed to invert the soil completely (Cunningham, 1967). Therefore take-all might be expected to be most severe after direct drilling, which leaves large root and stem base fragments (which provide the major source of inoculum near the soil surface. These expectations have been fulfilled in trials in the Pacific North West of the USA (Moore & Cook, 1978; Cook & Reis, this volume).

Redistribution of inoculum is only one of the ways in which cultivation treatment can influence the incidence of take-all. Hornby (1975), measured inoculum in the second year of a cultivation trial at Boxworth EHF on a site which had grown eight successive wheat crops and showed that there was, initially, more inoculum in the surface layers of unploughed than ploughed soil. Between February and July, however, the infectivity of debris in the top 10 cm declined more rapidly in direct-drilled than in ploughed plots, presumably reflecting greater activity of competitors and antagonists of the take-all fungus in the microbially rich upper layers of the undisturbed soil. Many of the early direct-drilling trials were carried out on sites not many years out of old pasture, and Shipton (1969) showed that, here too, direct drilling could increase microbial antagonism against *Gaeumannomyces graminis*, possibly because the treatments affected persistence of *Phialophora radicicola*.

In the original direct drilling trials at Jealott's Hill, direct-drilled crops usually developed less take-all than those on otherwise comparable ploughed areas (Brooks & Dawson, 1968). Similar decreases in take-all after direct drilling have also been reported from Scotland (Lockhart *et al.*, 1975) and Ireland (Cunningham & Fortune, 1978). Results from a large number of ADAS trials, mainly on old arable sites in the drier counties of Eastern England, were rather different (Table 5). Direct drilling sometimes decreased and sometimes increased disease but differences were generally too small to affect yield. In these trials the greatest increase in take-all as a result of direct drilling was in the third year at Rawreth. However, at this site the direct-drilled plots had become severely infested with couch grass, a weed frequently encouraged by direct drilling (Cussans, 1975), and the pathogen may have multiplied on the rhizomes of this grass. Furthermore direct drilling on the heavy London clay of this site caused structural deterioration of the soil down to 30 cm. By the third year, soil compaction was sufficient to reduce the yield of the direct-drilled plots and the weakened roots of plants in these plots may have been more susceptible to disease. In the following year, when the compaction problem had been corrected by sub-soiling, take-all in the direct-drilled and ploughed plots was similar.

Take-all was also most severe in the direct-drilled plots in the third year of the Tetworth trial, possibly because they were sown 10 weeks earlier than the others. Whether the increased take-all was due solely to earlier drilling is impossible to determine from these results. However, it is important to note that a major effect of non-ploughing husbandry is to reduce the autumn labour demand, which encourages early sowing. Any effect of sowing date on take-all is therefore important.

Table 5. Effects of cultivation on take-all incidence in three successive wheat crops

Site	Previous cropping	Treatment	% plants with take-all			
			1st year	2nd year	3rd year	Mean
Rawreth	Wheat after	MPB	74	4	26	35
Essex	Oats	RC	59	7	19	28
(CL, Windsor)*	(1974)†	DD	67	6	53	42
Hundon	Beans	MBP	27	31	24	27
Suffolk	(1973)	RC	22	27	23	24
(CL, Hanslope)		DD	33	25	24	27
Knapwell	3 Wheat	MBP	70	45	40	52
Cambs	crops	RC	74	62	34	57
(CL, Wicken)	(1973)	DD	67	51	41	53
Gt Easton	Wheat	MBP	92	59	41	64
Essex	(1973)	RC	90	60	47	66
(CL, Hanslope)		DD	92	33	42	53
Tetworth	Wheat	MBP	76	56	28	53
Hunts	(1973)	RC	80	47	35	54
(CL, Denchworth)		DD	67	49	43	53
Emneth	2 Wheat	MBP	0	20	24	15
Norfolk	crops	RC	0	22	26	16
(Org ZyCL)	(1976)	DD	0	16	36	17
		MBP	57	34	31	41
Mean		RC	54	38	31	41
		DD	54	30	40	41

* Soil type and series; CL, Clay Loam; Org ZyCL, Organic Silty Clay loam
† 1st year of trial.

Although there is no marked, consistent effect of cultivation method on take-all in England, the severity of the disease can often be linked with cultivation practice. Anything which restricts root growth, such as cultivation systems which cause deterioration in soil structure, will increase the disease severity. Such deterioration can result from direct drilling on London clay, or from ploughing under very wet conditions which leads to smearing of the plough sole or development of a plough pan. Under English conditions the severity of take-all is probably more dependent on the ability of the cultivator to overcome the soil structure problems than on methods he uses to do this.

OTHER DISEASES

Observations made by the ADAS Aerial Photography Unit suggest that direct drilling may aggravate barley stunt disorder, a condition frequently associated with (though not conclusively attributable to) root infection by certain strains of *Rhizoctonia solani* (McKelvie, 1978) In one field at Gleadthorpe EHF patches of "stunt" were observed in a barley crop in 1978. The patches reappeared in 1979 only in the direct drilled part of the field and were not seen where a mould-board plough had been used (T.S. Bell, personal communication). Observations in Scotland have also shown barley stunt to be worse in undisturbed than in disturbed soil (Anon., 1976).

Blackgrass can be favoured by direct drilling (Cussans, 1975) and since the strain of ergot (*Claviceps purpurea*) which infects blackgrass also infects wheat (Mantle *et al.*, 1977) interactions between cultivatio method, weed and disease are possible. Ergot sclerotia are not buried when crops are direct drilled and there have been occasional reports that ergot has become common on some farms where direct drilling is regularly practised.

Grass weeds were also among the factors investigated by Howell (1969) to explain the increased incidence of wheat leaf stripe (*Cephalosporium gramineum*) on direct-drilled fields. Direct drilling may also favour this disease indirectly through its effect on pests. After old grass, wireworm numbers can be two to three times greater on direct-drilled than on ploughed land (Edwards, 1975) and this could facilitate attack by *C. gramineum*, which enters the plant through damaged roots (Slope & Bardner, 1965).

CONCLUSIONS

It is obvious that the move away from mould-board ploughing can result in more pathogen inoculum being left on or near the soil surface. Effects of treatments on amounts of inoculum are, however, often masked by effects on crop growth and microclimate.

With take-all it is quality of cultivation rather than the method used which determines disease severity. The aim should be to use techniques which will, on any particular soil type, minimize the likelihood of root restriction caused by structural problems. With

barley stunt disorder the situation may be different and current,
limited information suggests that ploughing may be necessary to control
this disease.

For plant pathologists, the long-term importance of recent developments in cultivation techniques will probably not be their direct
effects on diseases but rather their indirect effects, because non-
ploughing husbandry encourages farmers to drill earlier. September-
sown crops are more likely than those sown later to become infected
by mildew, rusts and barley yellow dwarf virus in the autumn (Plumb,
this volume) and to develop more severe eyespot. The benefits of
early drilling are too great for these increased disease risks to argue
strongly against it but this trend may necessitate some rethinking of
our approach to disease control in the autumn.

REFERENCES

Anon. (1976) Barley Stunt Disorder. *North of Scotland College of
Agriculture Investigations and Field Trials* 1974-75, pp. 220-1.
Brooks D.H. & Dawson M.G. (1968) Influence of direct-drilling of
winter wheat on incidence of take-all and eyespot. *Annals of
Applied Biology* 61, 57-64.
Cox J. & Cock L.J. (1962) Survival of *Cercosporella herpotrichoides*
on naturally infected straws of wheat and barley. *Plant Pathology*
11, 65-6.
Cunningham P.C. (1967) A study of ploughing depth and foot and root
rots of spring wheat. *Irish Journal of Agricultural Research* 6,
33-9.
Cunningham P.C. & Fortune A. (1978) Influence of method of seed bed
preparation on take-all. *An Foras Taluntais. Plant Sciences and
Crop Husbandry Research Report* 1977, p.30.
Cussans G.W. (1975) Weed control in reduced cultivation and direct
drilling systems. *Outlook on Agriculture* 8, 240-2.
Davies D.B., Vaidyanathan L.V., Rule J.S. & Jarvis R.M. (1979) Effect
of sowing date and timing and level of nitrogen application to
direct drilled winter wheat. *Experimental Husbandry* 35, 122-31.
Edwards C.A. (1975) Effects of direct drilling on the soil fauna.
Outlook on Agriculture 8, 243-4.
Harris D. (1979) Survival of the glume blotch fungus on stubble &
straw residues. *Straw decay and its effect on disposal & utilisation.* (Ed. by E. Grossbard), pp. 21-28. Wiley & Sons, Chichester.
Hornby D. (1975) Inoculum of the take-all fungus: nature, measurement,
distribution and survival. *EPPO Bulletin* 5, 319-33.
Howell M.J. (1969) The appearance of leaf stripe (*Cephalosporium
gramineum*) in cereals grown under conditions of minimum cultivation.
Proceedings 5th British Insecticide and Fungicide Conference 1, 34-8.
Lockhart D.A.S, Heppel V.A. & Holmes J.C. (1975) Take-all (*Gaeumannomyces graminis* (Sacc.) Arx & Olivier) incidence in continuous
barley growing and effect of tillage method. *EPPO Bulletin* 5, 375-83.
Mantle P.G. Shaw S. & Doling D.A. (1977) Role of weed grasses in the
etiology of ergot disease in wheat. *Annals of Applied Biology* 86,

339-51.

McKelvie A. (1978) Barley stunt disorder. *North of Scotland College of Agriculture. College Digest for 1978*, pp. 45-50.

Moore K.J. & Cook R.J. (1978) Influence of no-tillage on take-all of wheat. *Abstracts of Papers, 3rd International Congress of Plant Pathology, Munich* 1978, p. 175.

Shipton P.J. (1969) Take-all decline. *Ph.D Thesis, University of Reading*.

Slope D.B. & Bardner R. (1965) Cephalosporium stripe of wheat and root damage by insects. *Plant Pathology* 14, 184-7.

Stedman O.J. (1977) Effect of paraquat on the number of spores of *Rhynchosporium secalis* on barley stubble and volunteers. *Plant Pathology* 26, 112-20.

Yarham D.J. & Hirst J.M. (1975) Diseases in reduced cultivation and direct drilling systems. *EPPO Bulletin* 5, 287-96.

Zadoks J.C., Chang T.T. & Zonzak C.F. (1974) A decimal code for the growth stages of cereals. *Weed Research* 14, 415-21.

Cultural control of soil-borne pathogens of wheat in the Pacific North-West of the U.S.A.

R. J. COOK & E. REIS*
Department of Plant Pathology, Washington State University,
Pullman, Washington 99164, U.S.A.

INTRODUCTION

In the Pacific Northwest of the U.S.A., 90-95% of the wheat is winter-sown and grown under one of four specialized systems (Table 1). Fifteen soilborne pathogens of wheat are now recognized as important: *Urocystis agropyri* (Purdy & Holton, 1963) and *Tilletia controversa* (Purdy *et al.*, 1963), causes of flag smut and dwarf bunt, respectively; *Typhula idahoensis*, *T. incarnata*, *T. ishikariensis* and *Fusarium nivale*, causes of snow mold (Bruehl *et al.*, 1966); *Pythium iwayamai* and *P. okano-genensis*, causes of snow rot (Lipps & Bruehl, 1978; Lipps, 1979); *P. ultimum* and *P. aristosporum*, causes of root rot (Lipps, 1979; Cook *et al.*, 1980); *Fusarium roseum* 'Culmorum' and 'Graminearum', causes of dryland crown and foot rot (Cook *et al.*, 1968); *Gaeumannomyces graminis*, the cause of take-all (Cook *et al.*, 1968); *Pseudocercosporella herpotrichoides* (= *Cercosporella herpotrichoides*), the cause of straw breaker foot rot (eyespot) (Bruehl *et al.*, 1968) and *Cephalosporium gramineum*, the cause of vascular stripe disease (Bruehl, 1956).

These pathogens have become increasingly important as crop management has become more intensive. Research programmes in the Northwest states (Idaho, Montana, Oregon, Washington and Utah) are investigating the ecology of these soilborne pathogens in the expectation that disease can be reduced by modification of cultural practices. This chapter only reports the progress on fusarium crown and foot rot, and take-all, since these represent the major part of our research programme.

FUSARIUM CROWN AND FOOT ROT

Severe fusarium crown and foot rot was first seen in central Washington in the 1960s (Cook, 1968a), in wheat after fallow. Its appearance was

---

* Present address: Centro Nacional de Pesquisa de Trigo, Caixa Postal, 569, 99.100 - Passo Fundo, RS, Brazil.

Jenkyn J. F. & Plumb R. T. (1981) *Strategies for the Control of Cereal Disease*

Table 1. Major wheat management systems* and accompanying soil-borne diseases in the Pacific Northwest

System	Annual precipitation	Features†	Most important diseases caused by soilborne pathogens
Wheat-fallow	45–50 cm	Wheat seeded early on stubble-mulched fallow	Pseudocercosporella and fusarium foot rots, snow mold, flag smut, snow rot, cephalosporium stripe
Annual-cropped, intermediate-rainfall area	between 45–50 cm and 70–80 cm	Late-seeded wheat in mono-culture or rotation with barley, dry peas or lentils	Snow mold and snow rot, pythium root rot, dwarf bunt, cephalosporium stripe and occasionally take-all
Annual-cropped, high-rainfall area	> 70–80 cm	Late-seeded wheat in rota-tion with a wide variety of crops, including veget-ables and small fruits	Take-all and occasionally pseudocercosporella foot rot
Irrigated, annual-cropped area	45–50 cm	Wheat seeded in any month between September and March in rotation with a wide variety of crops	Take-all and pseudocercosporella foot rot

* Numerous variations occur within each system. e.g. in date of seeding, rate of nitrogen, time of tillage, etc.

† The wheat-fallow system uses deep-furrow drills to seed wheat in rows 32-45 cm apart whereas the other three systems use variations of a disc drill to seed wheat in rows 20-25 cm apart.

related to severe water stress in semi-dwarf wheats given large amounts of nitrogen (Papendick & Cook, 1974).

Early efforts to control this disease attempted to reduce populations

of the pathogen.   Because oats favour multiplication of 'Culmorum' under
Northwest conditions (Cook, 1968b), they are now rarely grown in the
dryland wheat-fallow area.   Stubble mulching (straw left on, or mixed
into, the soil surface to reduce wind erosion) encourages the development of airborne saprophytes in the straw while it is still above-ground
and this prevents saprophytic colonization by 'Culmorum' once the straw
is buried (Cook & Bruehl, 1968; Cook, 1970).   Anhydrous ammonia used as
a "fumigant" also reduces populations of the pathogen but is ineffective
at rates commercially and agronomically acceptable (Smiley *et al.*, 1970,
1972).

Current strategy is to manage disease by manipulating water potential
(water stress) within the host.   Plants supplied with adequate water and
with midday water potentials above -32 to -35 bars (nonstressed) rarely,
if ever, develop severe fusarium crown and foot rot under Northwest
conditions.   Although the pathogen may colonize nonstressed plants it
causes little damage.   By contrast, spread of fusarium through wheat
plant tissues at water potentials below -32 to -35 bars (tissues under
stress) is aggressive and results in acute crown and foot rot and pre-
mature senescence.   Acute disease probably occurs when plant water
potentials are ideal for growth and metabolism of the pathogen (Cook
et al., 1972; Cook & Christen, 1976) but limit physiological resistance
in the host (Cook, 1973).

A delay of 7-10 days in sowing, i.e. from late August until early
September, and reductions in N applied from 90-120 kg/ha to 60-70 kg/ha
have been the two most effective means of decreasing total transpiration and prolonging the supply of soil water under Northwest conditions
(Papendick & Cook, 1974).   A well-maintained strawdust mulch to slow
evaporative loss of water from the soil (Papendick *et al.*, 1973) wide
row spacing (40 cm) and low seed rates (50 kg/ha) have also helped to
delay plant water stress.

Annual precipitation averages only 20-35 cm in areas where fusarium
crown and foot rot occurs.   Although each wheat crop is preceded by a
year of fallow, total water available for a given crop rarely exceeds
60% of the precipitation for the 2 years because of run-off and
evaporation.   The use of 60-70 kg N/ha is adequate under these conditions, as indicated by no less and, in some infested fields, 10-20%
more yield than with higher N rates.   The tendency to use excessive N
began in the 1960s when the lodging-resistant, semi-dwarf cultivar
Gaines became available.   Growers in the drier areas of the Northwest
attempted to obtain yields similar to those possible in wetter areas
but in so doing created the stress that favours fusarium crown and foot
rot.

Attempts are being made to select wheats that use water more slowly
and penetrate soil more deeply, or have some resistance to fusarium.
Test cultivars are sown after fallow in late August, in an area which
has 25 cm annual rainfall and where the 'Culmorum' population is about
1000 propagules/g of the surface 10 cm of soil; the disease is encouraged by applying 100-120 kg N/ha.   Cultivars that remain green longest

and/or yield the most are tested more thoroughly. The goal is to grow wheat that avoids or tolerates water stress (Cook, 1973) with management systems that conserve the limited supplies of soil water.

TAKE-ALL

Take-all in the Northwest occurs in the irrigation districts and west of the Cascade Mountains where soil conditions are favourable to *Gaeumannomyces graminis* var. *tritici* (Cook, 1980). Soils in the irrigation districts are alkaline (pH 7.0-7.5), well-drained, of sandy texture and have little organic matter. Most have been cropped for 20 years or less. Soils west of the Cascades are also well-drained and, although naturally near pH 5.0, most have been limed and are of pH 5.5 to 7.0. Both areas provide the high water potential necessary for growth of the pathogen (Cook *et al.*, 1972; Cook & Christen, 1976).

Wheat grown in rotation with potatoes, corn (maize), oats, beets or most other non-hosts does not develop severe take-all, provided successive crops of wheat or barley are not grown. West of the Cascades, take-all decline occurs as early as the third or fourth consecutive wheat crop (Baker & Cook, 1974) but it takes longer to develop in the irrigation districts. However, the interruption of consecutive wheat crops with rotation crops such as oats, potatoes, beans and especially alfalfa (lucerne) counteracts take-all decline (Cook, 1979) and this could account for severe take-all in wheat after alfalfa in South Africa (Louw, 1957) and Northwest U.S.A. (Cook *et al.*, 1968). Although alfalfa apparently makes soil more conducive to take-all, grass weeds in the crop may also allow *G. graminis* var. *tritici* to multiply.

The object of our research on take-all is to find management practices and soil treatments that will provide disease control. Garrett (1948) showed that plants can tolerate considerable root rot caused by *G. graminis* var. *tritici* in fertile soils because they are able to form new roots rapidly. Furthermore, Fellows (1938) showed that *G. graminis* var. *tritici* is unlikely to kill plants unless the crown is infected. Fumigation can be effective in the year of treatment but subsequently the disease may be even more severe than in untreated plots (Ebbels, 1969). Therefore, our strategy is to permit some root rot but use practices that prevent the pathogen from damaging the crown. In particular, we have been studying the effects of water management, tillage, form of nitrogen, phosphorus and trace element nutrition on disease, and ways to manage the take-all decline phenomenon.

Water management

Growth of *G. graminis* var. *tritici* is halved at about -20 bars water potential and virtually prevented below about -45 bars (Cook *et al.*, 1972). Under conditions of high evaporation, water potentials in the range -20 to -40 bars develop rapidly both in the tillage layer and within wheat plant tissues. Because the pathogen is ectotrophic,

spreading by runner hyphae from root to root and into the crown, controlled irrigation can greatly slow its spread. Take-all control is
greatest when the total water for a crop is split between only three or
four applications. Such a programme permits extensive drying of the
soil at the surface and around plant roots between applications and
limits take-all. Unfortunately in the Northwest labour-saving pivot
irrigation systems are now commonly used (about 230,000 ha of wheat was
irrigated by pivots in 1979). These apply about 1 cm of water every
12-24 hr which is ideal for take-all.

Tillage and crop residue management

In England, Brooks & Dawson (1968) found that take-all of wheat after
grass was less in plots which had been direct-drilled than in those
conventionally tilled but in Northwest U.S.A., take-all is favoured by
no-till (Moore, 1978). Moore did several experiments in eastern,
central and western Washington State with spring and winter wheat; in
all cases, both incidence and severity of take-all were higher and
yields lower in no-till than in tilled plots. In plots fumigated with
methyl bromide to eliminate take-all, yields were equally high with
either method of tillage. When oat inoculum of the take-all fungus was
introduced at sowing, take-all in fumigated plots was equally severe
whether the plots had been tilled or not. These and other results
indicate that the greater amount of take-all after no-till is the result
of more inoculum (infected residues, mainly crown tissues) in the no-
till plots and the fact that the inoculum is ideally positioned next to
the crowns of the direct-sown wheat. Where reduced tillage is important
in decreasing erosion in the Northwest, it may be necessary to accept
some take-all as inevitable.

Form of nitrogen

Ammonium nitrogen can decrease take-all under Northwest conditions
(Huber *et al.*, 1968) apparently because it decreases pH at the root
surface (Smiley & Cook, 1973). The effect has been demonstrated primarily with dry fertilizers (ammonium sulphate, ammonium phosphate and
ammonium chloride) broadcast and mixed into the soil. Nitrogen sta-
bilizers such as N-serve (2-chloro-6-(trichloromethyl) pyridine) sometimes improve control but results have not been consistent. Ammonium
seems to be more effective against take-all west of the Cascades than
in the irrigation districts to the east, perhaps because the soils on
the west side are more acid. Chloride ions also decrease take-all under
some conditions (R.L. Powelson & T. Jackson, personal communication)
although the mechanism is unknown. There has been some substitution of
ammonium chloride for other nitrogen fertilizers in the Willamette
Valley of Oregon.

Phosphorus and trace element nutrition

In both pot and field experiments, wheat plants given ample phosphorus
had more roots and fewer infected roots than plants deficient in
phosphorus. Thus phosphorus, like nitrogen (Garrett, 1948), increases

root development. However the decrease in numbers of infected roots per plant shows that phosphorus also enhances the resistance of root tissues to *G. graminis*, and confirms the observations of earlier workers that phosphorus decreases take-all.

Take-all on plants grown in silica sand in growth chambers was increased by deficiencies of zinc, copper, manganese or iron. Either copper or zinc sprayed onto the leaves of young wheat plants grown under these conditions gave significant control of take-all, which suggests that their effect was on the host and not directly fungicidal. Significant reductions in take-all also resulted in a yield response to zinc in one field trial, and in a response to a mixture of copper, zinc, manganese and iron in another.

Trace elements are most readily available to plants under acid conditions (Hausenbuiller, 1972; Lindsay, 1972; Lucas & Knezek, 1972). For example, an extractable zinc concentration of 6.5 ppm at pH 5.0 becomes only 0.007 ppb at pH 8.0. Possibly the greater amounts of take-all in limed or naturally alkaline soils are due to trace element deficiencies. In agreement with the results of Ponchet (1962), calcium had no direct effect on take-all in our experiments but was important because it increased soil pH.

Take-all decline

Take-all decline is well documented for many wheat-growing areas (Slope & Cox, 1964; Baker & Cook, 1974; Shipton, 1975; Cook & Rovira, 1976; Hornby, 1979; Rovira & Wildermuth, 1980; Prew, this volume). The following discussion is limited to our own results and their application in Pacific Northwest U.S.A.

Our field experiments have shown that the factor or factors responsible for take-all decline can be transferred from field to field (Shipton *et al.*, 1973; Baker & Cook, 1974) and become ineffective if monoculture wheat is interrupted with a break crop such as alfalfa, oats, beans or potatoes (Cook, 1979). Our results (Shipton *et al.*, 1973; Cook & Rovira, 1976) suggest that the responsible agent or agents are sensitive to moist heat (60°C for 30 min), methyl bromide and chloropicrin. These observations are consistent with findings for take-all decline in other areas (Shipton, 1975).

We are studying four different microbiological mechanisms which may be involved in take-all decline, namely (i) hypovirulence in the pathogen (Lapierre *et al.*, 1970); (ii) cross protection by fungi related to *G. graminis* var. *tritici* (Deacon, 1976; Wong & Southwell, 1979); (iii) antagonism by bacteria in the rhizosphere or on the rhizoplane (Cook & Rovira, 1976) and (iv) vampyrellid amoebae (Old, 1977; Anderson & Patrick, 1978).

Hypovirulence   After take-all decline has become established in a field, *G. graminis* var. *tritici* can be isolated from an average of 50-60% of randomly selected wheat plants. Of these isolates, about

90% have proved virulent on wheat seedlings grown in either vermiculite or fumigated soil. A similar percentage of ascospore cultures from these isolates have also proved virulent. The original field isolates commonly lose virulence after several months in the laboratory but this is probably different from the hypovirulence found by Lapierre *et al.* (1970). We have no evidence that hypovirulence is important in the field.

Cross protection   All of our isolates fit the description (Walker, 1972) of *G. graminis* var. *tritici*. *Phialophora graminicola* has not been reported in the U.S.A., although our method of isolation, where the pathogen is "baited" from the root system with one or two wheat seedlings, possibly selects in favour of *G. graminis* var. *tritici* and against nonpathogens. However, two isolates of *P. graminicola* supplied by J.W. Deacon, and two isolates each of *G. graminis* var. *graminis* and a *Phialophora*-like fungus supplied by P.T.W. Wong were tested against *G. graminis* var. *tritici* in one glasshouse and seven field trials. One isolate each of *P. graminicola*, the *Phialophora*-like fungus, and *G. graminis* var. *graminis* gave significant protection of wheat against *G. graminis* var. *tritici* in the glasshouse trial but none gave protection in the field against either natural or introduced inoculum of *G. graminis* var. *tritici*. *P. graminicola* is believed to be largely responsible for protection of wheat planted after grass leys in the United Kingdom (Deacon, 1973). Suppression of take-all was also observed in Washington State in wheat after grass (a mixture of smooth brome and intermediate wheat grass) (Cook, 1979) but we do not yet know whether *P. graminicola* or other avirulent fungi are involved.

Bacteria   Recent studies have suggested that bacteria in soil, or on wheat roots, especially fluorescent pseudomonads, may protect wheat against take-all (Cook & Rovira, 1976; Sivasithamparam *et al.*, 1978; Smiley, 1978; Campbell & Faull, 1979).

   Partial control of take-all on spring wheat has been obtained in field trials using non-sporing bacteria originally isolated from roots of wheat grown in decline soil (D.M. Weller, personal communication). Of about 35 different bacterial treatments applied to the seed, at least four decreased root infection and the numbers of whiteheads. One treatment was a combination of two fluorescent pseudomonads, two were fluorescent pseudomonads tested singly, and the fourth was a probable pseudomonad, but not *Ps. fluorescens*. In pots, the combination treatment gave almost complete control of take-all on seedlings. Whether any of these bacteria are responsible for take-all decline remains unknown.

Amoebae   Amoebae of the family Vampyrellidae, identified as *Arachnula impatiens* Cienk., occur in Washington soils where consecutive wheat crops have been grown and cause perforations in the pigmented hyphae of *G. graminis* var. *tritici* (Homma *et al.*, 1979). However, their activity is limited to soil drier than -10 to -100 mbar (i.e. drained sufficiently to permit good aeration) but not drier than -200 to -250 mbar (Cook & Homma, 1979). As amoebae are inactive in soils at saturation

(0 bars) on the one hand, and at field capacity (-300 mbar) or drier on
the other hand, they are unlikely to be implicated in take-all decline.

CONCLUSIONS

The occurrence of 15 important soilborne pathogens on wheat in Pacific
Northwest U.S.A. reflects the great diversity of environments and
cultural practices for wheat in this region. The two diseases discussed in detail in this paper reveal how cultural practices can be
used for disease control.

   Fusarium crown and foot rot can be controlled by practices that
reduce or delay plant water stress. The important practices are (1)
maintenance of a proper dust mulch during the summer fallow to prevent
loss of soil water by evaporation; (2) delaying seeding until mid-
September to shorten the period of water usage by the crop and (3)
application of less nitrogen to reduce leaf area and hence total transpiration. In addition, some progress has been made in the development
of cultivars that avoid or possibly tolerate water stress. This management of plant water status through cultural practices and choice of
cultivar has increased yields in fusarium-infested fields by 10-20%.

   Take-all in the irrigation districts can be controlled in consecutive wheat crops by (1) the use of sufficient tillage to fragment and
bury infected crowns; (2) irrigating the crop on only three or four
occasions during the growing season, so that the soil and critical
plant surfaces dry between applications; (3) use of ammonium nitrogen
to reduce rhizosphere pH (especially in western Washington and Oregon
where soil pH is commonly 6.0 or less) and (4) use of phosphorus and
trace element (especially zinc) fertilizers to enhance crown root
development and host resistance. Significant benefit is also derived
from take-all decline but only after the third or subsequent consecutive wheat crop. Break crops such as alfalfa, potatoes, oats and soya-
beans help to control take-all, presumably by starving the pathogen.
However, these crops also counteract take-all decline so that once the
pathogen is re-established, take-all is more severe than where wheat
has been grown continuously for many years. A factor capable of
initiating take-all decline has been transferred between field plots.
Hypovirulence, cross protection by related fungi and vampyrellid
amoebae are very unlikely to be the cause of take-all decline in the
Northwest. On the other hand, significant control of take-all has been
obtained in field plots by treating wheat seeds with fluorescent
pseudomonads originally isolated from roots of wheat grown in decline
soil. Treatment with bacteria may prove a practical means of take-all
control, especially after rotation crops where soils are highly conducive to this disease.

REFERENCES

Anderson R.R. & Patrick A.A. (1978) Mycophagous amoeboid organisms from soil that perforate spores of *Thielaviopsis basicola* and *Cochliobolus sativus*. *Phytopathology* 68, 1618-26.

Baker K.F. & Cook R.J. (1974) *Biological Control of Plant Pathogens*. W.H. Freeman and Co., San Francisco. 433 pp.

Brooks D.H. & Dawson M.G. (1968) Influence of direct-drilling of winter wheat on incidence of take-all and eyespot. *Annals of Applied Biology* 61, 57-64.

Bruehl G.W. (1956) Prematurity blight phase of *Cephalosporium* stripe disease of wheat. *Plant Disease Reporter* 40, 237-41.

Bruehl G.W., Nelson W.L., Koehler F. & Vogel O.A. (1968) Experiments with *Cercosporella* foot rot (straw breaker) disease of winter wheat. *Washington Agricultural Experiment Station Bulletin* 694. 14 pp.

Bruehl G.W., Sprague R., Fischer W.R., Nagamitsu M., Nelson W.L. & Vogel O.A. (1966) Snow molds of winter wheat in Washington. *Washington Agricultural Experiment Station Bulletin* 677. 21 pp.

Campbell R. & Faull J.L. (1979) Biological control of *Gaeumannomyces graminis*: field trials and the ultrastructure of the interaction between the fungus and a successful antagonistic bacterium. *Soil-borne Plant Pathogens* (Ed. by B. Schippers & W. Gams), pp. 603-9. Academic Press, London.

Cook R.J. (1968a) *Fusarium* root and foot rot of cereals in the Pacific Northwest. *Phytopathology* 58, 127-31.

Cook R.J. (1968b) Influence of oats on soil-borne populations of *Fusarium roseum* f. sp. *cerealis* 'Culmorum'. *Phytopathology* 58, 957-60.

Cook R.J. (1970) Factors affecting saprophytic colonization of wheat straw by *Fusarium roseum* f. sp. *cerealis* 'Culmorum'. *Phytopathology* 60, 1672-6.

Cook R.J. (1973) Influence of low plant and soil water potentials on diseases caused by soilborne fungi. *Phytopathology* 63, 451-8.

Cook R.J. (1979) Influence of crops besides wheat on *Gaeumannomyces graminis* and take-all decline. *Phytopathology* 69, 914.

Cook R.J. (1981) Effect of soil pH and physical conditions on take-all. *Biology and Control of Take-all* (Ed. by P.J. Shipton & M.J.C. Asher). Academic Press, New York. (In press.)

Cook R.J. & Bruehl G.W. (1968) Relative significance of parasitism versus saprophytism in colonization of wheat straw by *Fusarium roseum* 'Culmorum' in the field. *Phytopathology* 58, 306-8.

Cook R.J. & Christen A.A. (1976) Growth of cereal root-rot fungi as affected by temperature-water potential interactions. *Phytopathology* 66, 193-7.

Cook R.J. & Homma Y. (1979) Influence of water potential on activity of amoebae responsible for perforations of fungal spores. *Phytopathology* 69, 914.

Cook R.J., Huber D., Powelson R.L. & Bruehl G.W. (1968) Occurrence of take-all in wheat in the Pacific Northwest. *Plant Disease Reporter* 52, 717-8.

Cook R.J., Papendick R.I. & Griffin D.M. (1972) Growth of two root-rot fungi as affected by osmotic and matric water potentials. *Soil Science Society of America Proceedings* 36, 78-82.

Cook R.J. & Rovira A.D. (1976) The role of bacteria in the biological control of *Gaeumannomyces graminis* by suppressive soils. *Soil*

Biology and Biochemistry 8, 269-73.

Cook R.J., Sitton J.W. & Waldher J.T. (1980) Evidence for *Pythium* as a pathogen of direct-drilled wheat in the Pacific Northwest. *Plant Disease* 64, 102-3.

Deacon J.W. (1973) Control of the take-all fungus by grass leys in intensive cereal cropping. *Plant Pathology* 22, 88-94.

Deacon J.W. (1976) Biological control of the take-all fungus, *Gaeumannomyces graminis*, by *Phialophora radicicola* and similar fungi. *Soil Biology and Biochemistry* 8, 275-83.

Ebbels D.L. (1969) Effects of soil fumigation on disease incidence, growth, and yield of spring wheat. *Annals of Applied Biology* 63, 81-93.

Fellows H. (1938) Interrelation of take-all lesions on the crowns, culms, and roots of wheat plants. *Phytopathology* 28, 191-5.

Garrett S.D. (1948) Soil conditions and the take-all disease of wheat. IX. Interaction between host plant nutrition, disease escape, and disease resistance. *Annals of Applied Biology* 35, 14-7.

Hausenbuiller R.L. (1972) *Soil Science Principles and Practice*, Wm. C. Brown & Co., Dubuque, Iowa.

Homma Y., Sitton J.W., Cook R.J. & Old K.M. (1979) Perforation and destruction of pigmented hyphae of *Gaeumannomyces graminis* by vampyrellid amoebae from Pacific Northwest wheat field soils. *Phytopathology* 69, 1118-22.

Hornby D. (1979) Take-all decline: a theorist's paradise. *Soilborne Plant Pathogens* (Ed. by B. Schippers and W. Gams), pp. 133-56. Academic Press, London.

Huber D.M., Painter G.A., McKay H.C. & Peterson D.L. (1968) Effect of nitrogen fertilization on take-all of winter wheat. *Phytopathology* 58, 1470-2.

Lapierre H., Lemaire J.M., Jouan B. & Molin G. (1970) Mise en évidence de particules virales associées à une perte de pathogénicité chez le piétin-échaudage des céréales, *Ophiobolus graminis* Sacc. *Comptes Rendus des Séances de L'académie des Sciences (Paris)*, *Série D.* 271, 1833-6.

Lindsay W.L. (1972) Inorganic phase equilibria of micronutrients in soils. *Micronutrients in Agriculture* (Ed. by J.J. Mortved, P.M. Giordano & W.L. Lindsay), pp. 41-57. Soil Science Society of America, Madison, Wisconsin.

Lipps P.W. (1979) Etiology of snow rot of winter wheat. *Ph.D. Thesis, Washington State University, Pullman.*

Lipps P.E. & Bruehl G.W. (1978) Snow rot of winter wheat in Washington. *Phytopathology* 68, 1120-7.

Louw H.A. (1957) The effect of various crop rotations on the incidence of take-all (*Ophiobolus graminis* Sacc.) in wheat. *Department of Agriculture Union of South Africa Science Bulletin* 379. 12 pp.

Lucas R.E. & Knezek B.D. (1972) Climatic and soil conditions promoting micronutrient deficiencies in plants. *Micronutrients in Agriculture* (Ed. by J.J. Mortved, P.M. Giordano & W.L. Lindsay), pp. 265-87. Soil Science Society of America, Madison, Wisconsin.

Moore K.J. (1978) The influence of no-tillage on take-all of wheat. *Ph.D. Thesis, Washington State University, Pullman.*

Old K.M. (1977) Giant soil amoeba cause perforation of conidia of

Cochliobolus sativus. Transactions of the British Mycological Society 68, 277-81.

Papendick R.I. & Cook R.J. (1974) Plant water stress and development of *Fusarium* foot rot in wheat subjected to different cultural practices. *Phytopathology* 64, 358-63.

Papendick R.I., Lindstrom M.J. & Cochran V.L. (1973) Soil mulch effects on seedbed temperature and water during fallow in Eastern Washington. *Soil Science Society of America Proceedings* 37, 307-14.

Ponchet J. (1962) Étude des facteurs qui conditionnent le développement du piétin-échaudage: *Linocarpon cariceti* B. et Br. *Annales des Épiphyties* 13, 151-65.

Purdy L.H. & Holton C.S. (1963) Flag smut of wheat, its distribution and coexistence with stripe rust in the Pacific Northwest. *Plant Disease Reporter* 47, 516-8.

Purdy L.H., Kendrick E.L., Hoffmann J.A. & Holton C.S. (1963) Dwarf bunt of wheat. *Annual Review of Microbiology* 17, 199-222.

Rovira A.D. & Wildermuth G.B. (1981) The nature and mechanisms of suppression of take-all (*Gaeumannomyces graminis* var. *tritici*) in soil. *Biology and Control of Take-all* (Ed. by P.J. Shipton & M.J.C. Asher). Academic Press, New York. (In press.)

Shipton P.J. (1975) Take-all decline during cereal monoculture. *Biology and Control of Soil-borne Plant Pathogens* (Ed. by G.W. Bruehl), pp. 137-44. American Phytopathological Society, St. Paul, Minnesota.

Shipton P.J., Cook R.J. & Sitton J.W. (1973) Occurrence and transfer of a biological factor in soil that suppresses take-all of wheat in eastern Washington. *Phytopathology* 63, 511-7.

Sivasithamparam K., Parker C.A. & Edwards C.S. (1979) Bacterial antagonists to the take-all fungus and fluorescent pseudomonads in the rhizosphere of wheat. *Soil Biology and Biochemistry* 11, 161-5.

Slope D.B. & Cox J. (1964) Continuous wheat growing and the decline of take-all. *Report of the Rothamsted Experimental Station for 1963* p. 108.

Smiley R.W. (1978) Antagonists of *Gaeumannomyces graminis* from the rhizoplane of wheat in soils fertilized with ammonium- or nitrate-nitrogen. *Soil Biology and Biochemistry* 10, 169-74.

Smiley R.W. & Cook R.J. (1973) Relationship between take-all of wheat and rhizosphere pH in soils fertilized with ammonium vs. nitrate-nitrogen. *Phytopathology* 63, 882-90.

Smiley R.W., Cook R.J. & Papendick R.I. (1970) Anhydrous ammonia as a soil fungicide against *Fusarium* and fungicidal activity in the ammonia retention zone. *Phytopathology* 60, 1227-32.

Smiley R.W., Cook R.J. & Papendick R.I. (1972) *Fusarium* foot rot of wheat and peas as influenced by soil applications of anhydrous ammonia-potassium azide solutions. *Phytopathology* 62, 86-91.

Walker J. (1972) Type studies on *Gaeumannomyces graminis* and related fungi. *Transactions of the British Mycological Society* 58, 427-57.

Wong P.T.W. & Southwell R.J. (1979) Biological control of take-all in field trials using *Gaeumannomyces graminis* var. *graminis* and related fungi. *Soil-borne Plant Pathogens* (Ed. by B. Schippers and W. Gams), pp. 597-602. Academic Press, New York.

Fungicides, fertilizers and sowing date

J. F. JENKYN & M. E. FINNEY

Rothamsted Experimental Station, Harpenden, Hertfordshire

INTRODUCTION

Both crop nutrition and sowing date can affect the incidence and severity of cereal leaf diseases but modifying either is unlikely to be a practicable method of disease control. For example, most elements seem to significantly decrease powdery mildew (*Erysiphe graminis* DC.) only when supplied to plants deficient in that element (Jenkyn & Bainbridge, 1978). Nitrogen may be an exception because it usually increases susceptibility to most, if not all, cereal leaf diseases. Relatively small differences in amounts of nitrogen supplied, within the normal agricultural range, can have significant effects. However, even with nitrogen, disease control does not determine how much to apply, since large deviations from the nutritional optimum will decrease potential yield more than any increase in yield associated with a decrease in disease.

Sowing date also affects the severity of cereal leaf diseases but the effect is not the same for all pathogens and it is not usually important in deciding when crops are sown. As with crop nutrition, the yield penalties of deviating from the established optima can be large, and sowing date is often determined by factors such as weather and soil conditions, which are outside the grower's control.

Although there may be little scope for practical control of cereal leaf diseases by changing either the amount of nitrogen applied or the sowing date, the ways in which they affect diseases may nevertheless be important because they may interact with other control measures, especially the use of fungicides. An understanding of these interactions should make possible the more efficient use of fungicides and fertilizers to increase both the yield and quality of grain.

In this chapter the effects of nitrogen fertilizer and sowing date, and their interactions with fungicides, are illustrated by particular reference to powdery mildew of spring barley. Most of the nitrogen data were obtained from a series of factorial experiments done at

Rothamsted in 1975-1977 which tested all combinations of six nitrogen amounts, three nitrogen times (early, late and split equally between early and late), with and without mildew fungicide (tridemorph) and with and without rust fungicide (benodanil). Except where otherwise stated, effects of, and interactions between, factors are averaged over all other factors.

NITROGEN

Effects of nitrogen and fungicides on diseases and yield

Increasing the amount of nitrogen supplied usually leads to an increase in the severity of cereal powdery mildew (Last, 1962a). While this effect may be due in part to changes in crop density and hence micro-climate, Bainbridge (1974) also demonstrated that nitrogen increases growth and sporulation of *E. graminis*. However, nitrogen does not invariably increase mildew in crops, and at Rothamsted in 1970 it had the opposite effect (Jenkyn & Moffatt, 1975). In 1975 and 1976 early mildew was apparently increased by nitrogen up to 90 kg/ha but was dec-reased if greater amounts were used. We cannot yet explain these results but in all three years (1970, 1975 and 1976) the summers were dry and it may be that under these conditions large quantities of nitrogen do inhibit mildew.

When, as is usual, diseases are increased by extra nitrogen it is predictable that grain yields will often show interactions between nitrogen rates and fungicides (and also resistant varieties). Evidence for such interactions was obtained at Rothamsted in 1969 (Jenkyn & Moffatt, 1970) in experiments testing the effects of ethirimol on spring barley but many other experiments with cereal leaf pathogens have given similar results. The results shown in Table 1 are typical (see, for example, Dilz & Schepers, 1972; Ellen & Spiertz, 1975) and illustrate both the greater response to fungicide when much nitrogen is applied and the greater response to increasing amounts of nitrogen when mildew is controlled.

Table 1. Effects of nitrogen fertilizer and tridemorph on yield (t/ha) of spring barley (cv. Zephyr) (1975-1977)

	Nitrogen (kg/ha)					
	25	50	70	90	110	135
Without tridemorph	3.79	4.06	4.05	4.25	4.33	4.34
With tridemorph	4.14	4.38	4.72	4.99	4.92	4.96
SED	0.090					

Interactions between nitrogen fertilizers and fungicides may be predictable but the precise forms of these interactions are not. They may, nevertheless, be important as the amounts of nitrogen required to give maximum yield may depend upon whether or not fungicides are used. The average responses shown in Table 1 do not provide strong evidence that fungicide (tridemorph) affected either the amount of nitrogen needed for maximum yield or the economic optimum amount, but this is perhaps not surprising because they are averaged over years and over times of application. A more detailed examination of the data suggests that when all the nitrogen was applied to the seedbed, barley sprayed with fungicide responded up to the maximum tested (135 kg N/ha) whereas the yields of unsprayed plots were at a maximum with 110 kg N/ha.

In some instances, leaf diseases can affect the shape of the nitrogen response curve. In one series of experiments on spring barley at many sites, most responses were best described by two inter-secting straight lines (Boyd *et al.*, 1976); the exceptions were those sites affected by leaf diseases where smooth curves fitted the results better. However, results from a later series of experiments (Sparrow, 1979) did not show such an effect of leaf diseases on the shape of the response curve.

In a continuing series of trials done by the Agricultural Develop-ment and Advisory Service (ADAS) on winter barley, results from 29 trials (seven in 1977 and 22 in 1978) showed that use of fungicides commonly affected the shape of the nitrogen response curve but the size and form of this effect differed between sites. In many trials there was no conclusive evidence that fungicide had any effect on the economic optimum nitrogen rate. At those sites where there was good evidence for an interaction between nitrogen rate and fungicides, fungicides usually increased the optimum but by very different amounts in different trials (P. Needham, personal communication).

Interactions between nitrogen timing and fungicides can be expected to be particularly complex, and to differ from year to year, as nitrogen uptake is so dependent on weather. In experiments at Saxmundham (Suffolk) the best time to apply nitrogen, as measured by yield of spring barley, depends on rainfall in spring and summer. However, almost invariably, amounts of brown rust (*Puccinia hordei*) and the response to fungicide (benodanil) applied to control it are greater where nitrogen is applied as a top dressing than where it is applied to the seed-bed (Widdowson *et al.*, 1976).

In experiments with winter wheat in North Carolina applying nitrogen in spring instead of at sowing, markedly increased the damage done by mildew, measured by the yield response to fungicide application. Where mildew was not controlled, yields were similar whether nitrogen was applied to the seed-bed or as a top dressing in March but where mildew was controlled by fungicide, spring nitrogen gave better yields than seed-bed nitrogen (Hebert *et al.*, 1948).

At Rothamsted, applying nitrogen as a top dressing instead of to the seed-bed also tends to increase the severity of mildew on spring barley but effects have generally been smaller and less consistent than on brown rust (*Puccinia hordei*) at Saxmundham; perhaps because mildew commonly develops much earlier than does brown rust.

Effects of nitrogen and fungicides on grain quality

Grain quality is at present determined by size and nitrogen content although, in the future, protein composition may also become important.

Size (measured as 1000 grain weight (TGW)) is commonly increased by foliar fungicides (Jenkyn & Bainbridge, 1978) and sometimes decreased by nitrogen fertilizers. Although one might expect inter-actions between nitrogen rate and fungicides in determining grain weight, we have found little evidence for them, even in 1977 when nitrogen markedly decreased TGW.

The importance of nitrogen content depends on the intended use of the cereal. Sufficiently large nitrogen (protein) contents command a premium in the case of milling wheat but cereals intended for malting and subsequent use by the brewing industry (predominantly barley in the UK) are required to have a small nitrogen content. The nutritional value of cereals used for animal feed is theoretically affected by their nitrogen (protein) content although they are principally regarded as sources of metabolizable energy, and nitrogen content has no influence on their market value. However, an increase of only 1% (e.g. from 10 to 11%) in the protein content (corresponding to an increase of 0.16% in the nitrogen content) of the home-grown grain used each year in animal feeds by UK compounders is equivalent to a quantity of soya meal worth over £10M at current prices.

Control of leaf diseases might be expected to decrease nitrogen content of grain simply by the dilution effect of increasing grain size. However, measurements of grain nitrogen content in 1976 and 1977 provided evidence of interactions between nitrogen rate and fungicide (Table 2). When little nitrogen was applied, fungicide decreased nitrogen content, but when more was applied it either had no effect on nitrogen content (1976) or caused a substantial increase (1977; from 1.74 to 1.88% at 135 kg N/ha). Estimated amounts of nitrogen per 1000 grains confirm that mildew control can have a direct effect on actual amounts of nitrogen in grain when much nitrogen is used.

Our experiments have provided no evidence for interactions between time of nitrogen application and mildew fungicides in either grain size or grain nitrogen content.

Effects on the total amount of nitrogen harvested in grain

Nitrogen harvested in grain is the product of grain yield and grain nitrogen content and, not surprisingly, also shows marked interactions

Table 2. Effects of nitrogen fertilizer and tridemorph on nitrogen content (%) of barley grain (cv. Zephyr) in 1976 and 1977

| | Nitrogen (kg/ha) | | | | | |
	25	50	70	90	110	135
1976						
Without tridemorph	1.64	1.73	1.71	1.75	1.76	1.80
With tridemorph	1.49	1.64	1.65	1.70	1.77	1.79
SED	0.051					
1977						
Without tridemorph	1.59	1.69	1.72	1.67	1.77	1.74
With tridemorph	1.49	1.59	1.65	1.84	1.82	1.88
SED	0.043					

between the amount of applied nitrogen and fungicide. In our experiments the average increases in nitrogen removal due to mildew control by tridemorph were 13 and 9% in 1976 and 1977 respectively. The fungicide had little effect at small nitrogen rates but, in plots given 90 kg N/ha, the increases in nitrogen removed (= protein yield) were more than 20% in both years (Table 3).

Effects of disease on uptake and utilization of nitrogen

Healthy crops respond much better to, and make more efficient use of, applied nitrogen than do diseased crops but at present we have only a poor understanding of how leaf diseases affect the total nitrogen balance of a crop.

Diseased crops produce less dry matter than healthy ones and so might be expected to have a smaller total nitrogen demand but it does not follow that less need be applied. The root systems of plants affected by powdery mildew, for example, can be very much smaller than those of healthy plants (Last, 1962b; Paulech, 1969), and so presumably explore less soil and are less efficient in using nitrogen moved down the soil profile by rain.

Accelerated leaf senescence in diseased crops (Last, 1962b; Large & Doling, 1962; Brooks, 1972; Simkin & Wheeler, 1974; Jenkyn, 1976a)

Table 3. Changes (%) in amounts of nitrogen removed in barley
grain (cv. Zephyr), by applying tridemorph, at each nitrogen rate

| | Nitrogen (kg/ha) | | | | | |
	25	50	70	90	110	135
1976	+1.1	+8.6	+15.8	+21.6	+18.3	+11.4
1977	-2.3	-1.4	+ 4.1	+21.3	+12.7	+19.5

leading to a decreased supply of photosynthate may, however, affect
utilization of the nitrogen which is taken up from the soil. If
photosynthesis is relatively more adversely affected than nitrogen
uptake, so that there is an inadequate supply of photosynthate, then
nitrate will accumulate, especially in stem tissue (Darwinkel, 1975).
Such accumulation of nitrate as a result of mildew infection has been
demonstrated in pot experiments (Jenkyn, 1977), although root size is
probably less critical in pots than in fields. However, we now have
good evidence that similar effects do occur in crops (Table 4).

Table 4. Effects of nitrogen fertilizer and tridemorph on nitrate
nitrogen (µg/g dry weight) in shoots of barley (cv. Zephyr)
(27 June 1978)

| | Nitrogen (kg/ha) | | | | | |
	25	50	70	90	110	135
Without tridemorph	91.7	144.7	160.7	357.0	497.7	646.0
With tridemorph	85.7	69.3	116.3	204.7	289.0	399.7

In healthy plants a considerable proportion of the assimilated
nitrogen is recycled from the older, senescing leaves to younger
leaves and, ultimately, to the ear (Williams, 1955) but infection by
E. graminis, and probably by many other leaf pathogens, apparently
decreases such recycling (Finney, 1979). The protein content of a
senescing, mildewed leaf declines with time, as does that of a
senescing uninfected leaf. However, in the mildewed leaf, there is
a marked accumulation of amino-acids implying decreased recycling,
less efficient use of nitrogen and hence a greater total demand for
nitrogen in proportion to the amount of nitrogen harvested in grain.

We must also assume that some of the assimilated nitrogen is transferred to the pathogen and much of this subsequently dispersed in
spores. However, it is not known whether this represents a significant
drain on the total nitrogen resources of the crop or what proportion of
that lost in spores is recycled by their deposition on the soil within
the crop.

SOWING DATE

Although late sowing may discourage the development of leaf blotch of
spring barley (Lester, 1966) it usually favours cereal powdery mildew
(Last, 1955, 1957). Winter cereals similarly develop much less mildew
during the adult plant stages when sown early although they are at
much greater risk from mildew during the seedling stages (Jenkyn,
1976b). It can, therefore, be expected that mildew will be relatively
more damaging to late-sown crops than to early-sown and that there
will be interactions between sowing date and mildew fungicides (Last,
1955, 1957; Mórász, 1970). In recent experiments with spring barley
at Rothamsted (1976-79), interactions were detected in both 1977 and
1979 but not in 1976 and 1978 (Table 5). However, in 1976 the late-
sown plots were severely affected by the unusually dry conditions and
in 1978 mildew control was less good than in the other years.

Table 5. Effects of mildew fungicide on yield (t/ha) of early-
and late-sown spring barley

| | Year | | | |
	1976	1977	1978	1979
Yield of early-sown without fungicide	4.82	5.84	5.50	5.53
Increase with fungicide*	0.65	0.03	0.64	0.07
	(13.5%)	(0.5%)	(11.6%)	(1.3%)
Yield of late-sown without fungicide	3.06	5.79	5.01	4.25
Increase with fungicide*	0.49	0.80	-0.19	0.72
	(16.0%)	(13.8%)	(-3.8%)	(16.9%)

* Effects of ethirimol seed treatment plus tridemorph sprays in
1976 and 1977; ethirimol seed treatment in 1978; tridemorph
sprays in 1979.

The effect of sowing date on mildew development has not been
adequately explained although Last (1957) drew parallels with the
effects of nitrogen and suggested that both operated through changes
in growth rate. Recent studies by N. White at Rothamsted on the

effects of sowing date now lead us to believe that sowing date has little or no effect on susceptibility *per se*. In 1978, for example, spring barley (cv. Wing) sown on 8 March had nearly 80% less mildew over the whole season than did the same cultivar sown on 24 April but infection rates (Vanderplank, 1963), calculated for whole plants, differed surprisingly little for the two sowing dates.

Our evidence suggests that differences in amounts of mildew on crops sown at different times probably result simply from an interaction between the greater resistance of later-formed (i.e. upper) leaves and increasing "infection potential" (a combination of more favourable weather and greater amounts of inoculum) as the season progresses. Thus a late-sown crop generates a more severe epidemic because at any time it has earlier-formed (i.e. lower) and usually more susceptible leaves available for infection than does an earlier sown crop. If a very simple model is used which assumes that leaves are produced at a constant rate (say 1/week), that susceptibility of successively formed leaves steadily declines, and that "infection potential" shows a steady increase, then it is predictable that infection rates calculated for comparable individual leaves will be much greater for a late-sown than for an early-sown crop. Furthermore, the maximum leaf infection rate will occur on lower leaves in the late-sown than in the early-sown crop. Hypothetical curves based on this simple model closely resemble comparable curves based on field measurements on individual leaves in 1978 (N. White, personal communication).

CONCLUSIONS

Clearly, potentially high-yielding crops given much nitrogen are most at risk from cereal leaf diseases and, therefore, more likely to give economic responses to fungicides than are low-yielding, poorly-fertilized crops. Equally, crops protected from diseases will show a greater response to applied nitrogen and therefore use the applied fertilizer more efficiently, than will unprotected crops. Grain quality, in terms of both size and nitrogen content, is also likely to be improved if diseases are controlled.

Although such *generalizations* are possible, our understanding of the processes involved is imperfect so that we are unable to predict how the use of fungicides will affect nitrogen requirement or how these interactions will be affected by other conditions of soil or site. Frequently the effect of disease control on nitrogen requirement would appear to be small, at least in relation to the accuracy with which nitrogen requirement is usually determined. Greater efficiency in the use of nitrogen fertilizers may be achieved in the future but only when we understand better how diseases and other factors affect nitrogen uptake and assimilation and when techniques for measuring soil nitrogen are improved.

Generally cereal leaf diseases seem to be more favoured the later nitrogen is applied in the spring but there may nevertheless be advantages in delaying application of some or all of the nitrogen in order to decrease leaching losses or to increase the nitrogen (protein) content of grain. Crops so treated will be at particular risk from leaf diseases and most likely to respond to control measures.

Similarly, late-sown crops are more likely to be damaged by mildew, and several other leaf pathogens (Wiese & Ravenscroft, 1976), than are early-sown, and therefore more likely to need fungicide treatment. Maximum use should be made of genetic resistance, whenever the crop is sown, but it seems that where late sowing is unavoidable, cultivars with adult plant resistance will be most suitable.

REFERENCES

Bainbridge A. (1974) Effect of nitrogen nutrition of the host on barley powdery mildew. *Plant Pathology* 23, 160-1.
Boyd D.A., Yuen L.T.K. & Needham P.   (1976) Nitrogen requirement of cereals. 1. Response curves. *Journal of Agricultural Science, Cambridge* 87, 149-62.
Brooks D.H. (1972) Observations on the effects of mildew, *Erysiphe graminis*, on growth of spring and winter barley. *Annals of Applied Biology* 70, 149-56.
Darwinkel A. (1975) Aspects of assimilation and accumulation of nitrate in some cultivated plants. *Agricultural Research Reports (Wageningen)* 843, 1-64.
Dilz K. & Schepers J.H. (1972) Stikstofbemesting van granen. 24. Effekt van ziektebestrijding op de stikstofreaktie van winter- en zomertarwe met en zonder toepassing van chloormequat. *Stikstof* 71, 452-8.
Ellen J. & Spiertz J.H.J. (1975) The influence of nitrogen and benlate on leaf-area duration, grain growth and pattern of N-, P- and K-uptake of winter wheat *(Triticum aestivum)*. *Zeitschrift für Acker- und Pflanzenbau* 141, 231-9.
Finney M.E. (1979) The influence of infection by *Erysiphe graminis* DC. on the senescence of the first leaf of barley. *Physiological Plant Pathology* 14, 31-6.
Hebert T.T., Rankin W.H. & Middleton G.K. (1948) Interaction of nitrogen fertilization and powdery mildew on yield of wheat. *Phytopathology* 38, 569-70.
Jenkyn J.F. (1976a) Effects of mildew *(Erysiphe graminis)* on green leaf area of Zephyr spring barley, 1973. *Annals of Applied Biology* 82, 485-8.
Jenkyn J.F. (1976b) Observations on mildew development in winter cereals: 1968-73. *Plant Pathology* 25, 34-43.
Jenkyn J.F. (1977) Nitrogen and leaf diseases of spring barley. *Fertilizer Use and Plant Health, Proceedings of the 12th Colloquium of the International Potash Institute, Izmir, Turkey*, pp. 119-28.
Jenkyn J.F. & Bainbridge A. (1978) Biology and pathology of cereal powdery mildews. *The Powdery Mildews* (Ed. by D.M. Spencer),

pp. 283-321. Academic Press, London.

Jenkyn J.F. & Moffatt J.R. (1970) *Report of the Rothamsted Experimental Station for 1969, Part 1*, p. 152.

Jenkyn J.F. & Moffatt J.R. (1975) The effect of ethirimol seed dressings on yield of spring barley grown with different amounts of nitrogen fertilizer 1969-71. *Plant Pathology* 24, 16-21.

Large E.C. & Doling D.A. (1962) The measurement of cereal mildew and its effect on yield. *Plant Pathology* 11, 47-57.

Last F.T. (1955) Effect of powdery mildew on the yield of spring-sown barley. *Plant Pathology* 4, 22-4.

Last F.T. (1957) The effect of date of sowing on the incidence of powdery mildew on spring-sown cereals. *Annals of Applied Biology* 45, 1-10.

Last F.T. (1962a) Effects of nutrition on the incidence of barley powdery mildew. *Plant Pathology* 11, 133-5.

Last F.T. (1962b) Analysis of the effects of *Erysiphe graminis* DC. on the growth of barley. *Annals of Botany* 26, 279-89.

Lester E. (1966) Cereal diseases. 1. Leaf blotch of barley. *The Institute of Corn and Agricultural Merchants' Journal* 14, 16-7.

Mórász S. (1970) (Yield response of spring barley to powdery mildew infection when sown at different times.) *Növenytermelés* 19, 215-22. (Abst. in *Review of Plant Pathology* 50, 635).

Paulech C. (1969) Einfluss des Getreidemehltaupilzes *Erysiphe graminis* DC. auf die Trockensubstanzmenge und auf das Wachstum der vegetativen Pflanzenorgane. *Biológia, Bratislava* 24, 709-19. (Abst. in *Review of Plant Pathology* 49, 1619).

Simkin M.B. & Wheeler B.E.J. (1974) Effects of dual infections of *Puccinia hordei* and *Erysiphe graminis* on barley, cv. Zephyr. *Annals of Applied Biology* 78, 237-50.

Sparrow P.E. (1979) Nitrogen response curves of spring barley. *Journal of Agricultural Science, Cambridge* 92, 307-17.

Vanderplank J.E. (1963) *Plant Diseases: Epidemics and Control.* Academic Press, London. 349 pp.

Widdowson F.V., Jenkyn J.F. & Penny A. (1976) Results from two barley experiments at Saxmundham, Suffolk, measuring effects of the fungicide benodanil on three varieties, given three amounts of nitrogen at two times 1973-4. *Journal of Agricultural Science, Cambridge* 86, 271-80.

Wiese M.V. & Ravenscroft A.V. (1976) Planting date affects disease development, crop vigor, and yield of Michigan winter wheat. *Michigan State University, Agricultural Experiment Station, Research Report*, 314 pp.

Williams R.F. (1955) Redistribution of mineral elements during development. *Annual Review of Plant Physiology* 6, 25-42.

Disease management in high-input cereal growing in Schleswig-Holstein

H. C. EFFLAND
BASF AG, Holstenstrasse 88, 23 Kiel,
Federal Republic of Germany

INTRODUCTION

Inevitably disease management is a compromise. The costs and benefits of treatments intended to increase yield potential must be set against the costs and benefits of the consequent measures needed to protect that potential. Only long-term trials can determine where the balance lies, but an essential requirement for effective disease management is a well-educated farmer prepared to react quickly to changing disease risks using the best combination of available countermeasures. The intensive production techniques currently used for winter wheat, winter barley and winter oilseed rape in Schleswig-Holstein were developed after much experimentation and are adjusted annually to incorporate the best available information. Disease management is an integral part of these systems. All disease risks are carefully considered and the profits expected from their control calculated. Routinely applied control measures are avoided wherever possible.

CULTURAL MEASURES USED IN DISEASE MANAGEMENT

The use of intensive production techniques, which may be both expensive and labour consuming, only makes sense if they *consistently* produce large yields and profits (Pearson, this volume). To achieve such consistency, husbandry measures to safeguard this high yield potential are essential. The cereal plants must not be subjected to stress at any time from sowing until harvest, as even short term disturbances of plant growth can greatly increase the risk that these high-performance plants will be damaged by pathogens.

Nutrition must not be limiting and soil pH must be at the optimum throughout the growth period. Periods when nitrogen, especially, is either deficient or in excess must be avoided. Therefore it is now common to apply nitrogen 5-7 times instead of the 2-3 times usual in the past. Trace elements are applied prophylactically.

Nutrient uptake is facilitated by deep ploughing and good seed-bed

Jenkyn J. F. & Plumb R. T. (1981) *Strategies for the Control of Cereal Disease*

preparation so that strong roots develop very rapidly after planting and penetrate deeply into the soil. Rapid early growth ensures that plants pass quickly through the disease-susceptible germination and early growth phases and develop resistance in time to face the stresses of winter. A well-established plant also begins spring growth earlier. Good root growth is also encouraged by a generous supply of organic matter (achieved by straw incorporation). The inclusion of oilseed rape in a 3-year rotation improves the tilth and acts as a break to prevent build-up of pathogens which survive in soil or on crop debris.

If maximum yields are to be achieved early sowing is essential but it does increase the risk of damage by diseases. To counteract these risks basic or certified seed, treated against seed-borne diseases, should be used. With appropriate fungicidal support throughout their growth, early-sown cereals invariably produce large yields.

Optimal timing is as important as accuracy and precision in applying treatments to the crop, and adequate tractor capacity, tramlines and efficient machinery for applying fertilizers and sprays all contribute to this.

USE OF FUNGICIDES IN DISEASE MANAGEMENT

Intensive cereal production in Schleswig-Holstein involves use of large amounts of nitrogen (> 200-250 kg N/ha) and other practices including early sowing, dependence on few cultivars, and high plant and ear densities, which might be regarded as "mortal sins". We prefer to think of them as "calculated risks", which are accepted in the interests of greater yield potential and need to be matched by appropriate disease control.

Control of eyespot

Inoculum of *Pseudocercosporella herpotrichoides* (eyespot) can increase rapidly in short rotations, while infection is favoured by early sowing and by the favourable microclimate in dense crops. Under these conditions and with large amounts of nitrogen, lodging is clearly a risk. Our aim is to produce 6-7.5 million ears/ha on about 4.2-4.6 million plants, giving a yield of 8-9 t/ha. Thus even in dry weather the weight of grain per m^2 is 0.8-0.9 kg and in wet weather this can increase to 1.3 kg or more.

The risk of lodging is decreased by using stiff-strawed cultivars with good resistance to stem base diseases, providing they have the necessary genetic high yield potential. Chlormequat chloride + choline chloride ("5C", a growth regulator and fungistat) is applied to winter wheat two or three times between growth stages (GS : Zadoks *et al.*, 1974) 25-32. Winter barley is treated with the growth regulator mepiquat chloride + ethephon ("Terpal") at GS 32.

The control of eyespot relies to a large extent on the use of carbendazim-generating fungicides, applied between GS 30-32 and usually mixed with the second or third application of "5C" or "Terpal". These treatments are considered essential and applied routinely. Increasingly other fungicides (e.g. triadimefon, tridemorph, captafol or a thiocar-bamate) are added to this mixture to increase its spectrum of activity and achieve some control of other diseases (Fig. 1).

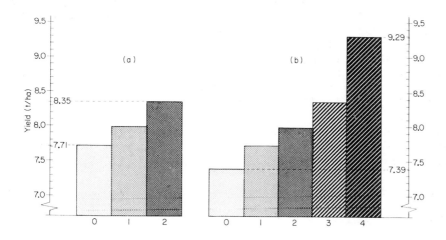

Figure 1. Average responses of (a) winter barley (17 experiments) and (b) winter wheat (11 experiments) to different pesticide regimes. (0 = untreated; 1 = carbendazim-generating fungicide at GS 30-32; 2 = fungicide mixture at GS 30-32; 3 = fungicide mixture at GS 30-32 and GS 52-58; 4 = fungicide mixture at GS 30-32, fungicide plus insecticide mixture at GS 39, fungicide plus insecticide plus trace element mixture at GS 52-58.)

Control of leaf and ear diseases

Epidemics of yellow rust, brown rust and mildew do not occur every year but when they do must be effectively controlled if the potential of this intensive system is to be achieved. Damaging attacks of *Septoria tritici* and *S. nodorum* on wheat, and *Rhynchosporium secalis* on barley also occur in some years.

In wheat ear diseases are considered more important than leaf diseases and most farmers control them prophylactically, applying fungi-cide mixtures at or before flowering. Damage by insects (e.g. aphids, thrips and midges) is prevented, where necessary, by applying insecti-cides at the same growth stage. There is no benefit from applying them later. The fear is not only direct insect damage, but also the develop-ment of saproprophytes as well as *Alternaria* spp. and *Cladosporium* spp. on honeydew. Figure 1 illustrates the average response of winter wheat in eleven experiments to different pesticide regimes.

In winter barley it is necessary to control diseases on the young plants in late autumn as well as in the spring. The spring fungicide, applied in a tankmix with the growth regulator, is intended to control eyespot, mildew, brown rust, yellow rust and rhynchosporium. In recent experiments fungicides applied at later growth stages (GS 55-69) to control diseases on the upper part of the stem and ears (rusts, mildew and septoria) have also given yield increases, in part by decreasing stem breaking.

Forecasts of damaging epidemics of pathogens not controlled prophylactically are based on crop observations. These can be effective over large areas because it is easy to survey the few cultivars grown. The cultivars most susceptible to each pathogen are used to forecast that disease on all cultivars. Crops of the "forecast cultivar" are treated routinely with fungicides as soon as the first symptoms of infection are seen.

EVALUATION OF CULTIVARS

The most important criteria in the selection of cultivars to grow in Schleswig-Holstein are yield potential and baking quality (winter wheat) or feeding and malting quality (winter barley). Each cultivar is tested under three management intensities. The genetically determined yield potential of each cultivar only becomes apparent when it is grown with high inputs (e.g. generous nutrition) and chemicals to decrease or eliminate weaknesses such as susceptibility to disease and lodging. Resistance to the main pathogens, though highly desirable, and genetically determined standing power are of secondary importance to high yield potential. Our experience shows that these weaknesses can be adequately countered by using crop protection measures.

Results from cultivar trials at Seegalendorf from 1966 to 1978 illustrate some of these points. Among the cultivars tested were Diplomat (susceptible to yellow rust on leaves and ears) and Caribo (susceptible to leaf mildew) both of which were grown every year. In the period 1966-73 both cultivars yielded more than the average of the other 12-15 cultivars tested. The yield of cv. Caribo was on average only 4% larger than that of cv. Diplomat. However, with increased inputs in the period 1974 to 1978 the yield of cv. Diplomat fell below the average of the other cultivars tested while cv. Caribo yielded considerably more. Over the 4 year period cv. Caribo yielded 10% more than cv. Diplomat. If there had been no high input tests the superiority of cv. Caribo would have been underestimated. The 1977 and 1978 trials used what is currently regarded as the best combination of inputs and cv. Caribo yielded 12.5 t/ha, exceeding the average of all new cultivars under test. This is remarkable for a cultivar that has been grown successfully for over 14 years. These trials also showed clearly that the genetic yield potential of cv. Diplomat (also grown since 1966) is inadequate under intensive management.

Similar comparisons were made between one feed wheat (cv. Maris

Huntsman) and five quality wheats (cvs Kobold, Kormoran, Monopol, Topfit, Vuka) grown at the same three management intensities. The yield of cv. Maris Huntsman is close to its maximum genetic potential with only medium inputs (Table 1) and is then greater than those of quality wheats; expensive crop protection measures of a high input system are unnecessary because response to them is small. The best of the other cultivars (*viz.* cvs Vuka and Kobold) show a much greater response to the crop protection measures of the high input system and only approach their genetic potential when these are used (Table 1). With high inputs yields of cvs Vuka, Kobold and Maris Huntsman are similar but growing quality wheats with high inputs is most profitable because more is paid for them.

Table 1. Yields (t/ha) of three winter wheat cultivars with different genetic yield potentials, grown with low, medium and high inputs. (Means of 10 trials done in 1977 and 1978 by the Chamber of Agriculture, Schleswig-Holstein).

Cultivar	Yield (t/ha)		
	Low input	Medium input	High input
Maris Huntsman	8.9	10.0	10.5
Vuka	8.3	9.1	10.5
Kobold	9.1	9.4	10.6

EFFECTS OF SOIL TYPE

The performance of cvs Diplomat and Caribo under different management schemes was tested on either sandy-loam or alluvial soils. With low inputs (over the period 1973 to 1975) it was only on alluvial soils that cv. Caribo yielded significantly more than cv. Diplomat. With high inputs (1977/78) the superiority of cv. Caribo was evident on both sandy-loam and alluvial soils.

ARE HIGH INPUTS PROFITABLE?

There is no absolute definition of high input cereal growing, indeed it is a very flexible concept and will be modified as techniques and economic circumstances change.

With current prices and costs intensively grown winter wheat or winter barley is clearly profitable on arable farms in Schleswig-Holstein if yields average more than 6 t/ha of good quality grain (Table

2). Large yields are directly related to large profits. The additional production costs needed to achieve these large yields are comfortably exceeded by the value of the extra output (Table 3). Plant protection accounts for 10% of the total variable and fixed costs, but the cost of fungicides, including application costs, represents only 2.5-3% of the total.

Table 2. Economics of cereal growing in Schleswig-Holstein.

	W-wheat		W-barley	
Yield (t/ha)	6.0	8.0	6.0	7.5
Output (£/ha)	780.00	960.00	636.00	795.00
Variable costs*				
Seed	47.00		44.00	
Fertilizers	110.00		97.00	
Plant protection	65.00		45.00	
Machine costs	47.00		44.00	
Others	56.00		52.00	
Total (£/ha)	325.00	352.00	282.00	308.00
Gross margin (£/ha)	455.00	608.00	354.00	487.00

* Fixed costs estimated at £275/ha.

To measure the profitability of a single input is difficult because inputs cost relatively less as output increases in response to increasing use of other inputs. Nevertheless we are confident that the direct benefits of pathogen control are considerable. Even if not, use of fungicides would still be essential to safeguard the increased potential resulting from other inputs.

CONCLUSIONS

Disease management is a small, but nevertheless vital, part of intensive cereal growing. Our experience shows that it is only successful when a flexible approach based on sound scientific knowledge is adopted. Treatments can then be matched to the potential of each cultivar at each location. Inflexible crop production systems and unplanned use of crop protection chemicals are no longer acceptable.

Poor harvests and the large yield fluctuations so common in the past

Table 3. The costs and benefits of growing winter wheat with different inputs. (Means of five milling quality cultivars, grown in 10 trials in 1977 and 1978 by the Chamber of Agriculture, Schleswig-Holstein.)

Management system	Yield (t/ha)	Increase in cost compared to low input (£/ha)	Increase in profit compared to low input (£/ha)
Low input*	8.50	-	-
Medium input†	9.35	27.00	70.00
High input¶	10.25	74.00	132.00

* 140 kg N/ha, growth regulator ("5C") applied once.
† 150–180 kg N/ha, "5C" applied once, fungicides to control foot diseases and mildew.
¶ 180–210 kg N/ha, "5C" applied twice, trace elements twice, fungicides to control foot, leaf and ear diseases and insecticides.

are now largely avoidable. High input cereal growing is undoubtedly very profitable if crops are grown with the same care and knowledge as speciality crops such as fruit and vegetables.

REFERENCES

Zadoks J.C., Chang T.T. & Konzak C.F. (1974) A decimal code for the growth stages of cereals. *Weed Research* 14, 415–21.

Constraints on cereal production and strategies of disease control

S. A. EVANS

Agricultural Development and Advisory Service,
Lawnswood, Leeds

MAXIMUM POTENTIAL YIELD

Maximum potential yield is a concept that has been much discussed over the last decade and has been estimated for a number of annual field crops in the United Kingdom. These estimates assume that all factors which influence crop growth are optimal, other than light, carbon dioxide and temperature which are not alterable. For cereals, potatoes and sugarbeet estimated potential yields are approximately two to three times current average farm yields (Alcock, 1967; Austin, 1978).

There have been some attempts to translate this concept into practice; the one receiving the most attention being the so called "potato blue-print". However, consistently achieving yields close to the estimated potential, even if possible, may not be currently practical or economic. Nevertheless, attempts to achieve maximum potential yields are useful as they can help to identify constraints and encourage the development of practices designed to remove them (Evans, 1977).

Consistently very high yields are not difficult to achieve with vegetative crops like potatoes and sugarbeet. This is not true for cereals; some very high yields have been recorded but the same treatment of crops elsewhere or in other seasons does not necessarily produce the same yields. Jenkins (1978) stated that it does not yet appear possible to suggest, for cereal crops, the inputs necessary to ensure a yield near to the crop's potential.

Those who have attempted to identify constraints on cereal yields (Fiddian, 1977; Austin, 1978) mostly agree that season, site, crop nutrition and cultivar are more important than diseases, pests and weeds. Site, which includes weather and soil, is clearly important in determining yield. The agriculturalist has no control over the former but, at least in theory, it ought to be possible to learn enough about effects of soil conditions to overcome any constraints they might impose. However, by using the range of techniques and fungicides already available it is almost possible to eliminate effects of diseases

in crops. If disease was the main factor preventing the achievement of
potential yield then those yields would be realised much more often.

We are left with the problem of explaining why cereal crops, adequately supplied with water and nutrients, and with losses caused by
diseases, pests and weeds prevented, frequently fail to reach their
potential.

FARM YIELDS

The farmer is not necessarily interested in maximum potential yield.
He wants the maximum economic return (Pearson, this volume). In an
established cereal growing system he has to assess the cost/benefit
ratio for any extra or changed operation. Pesticide use has become
widespread because it is generally very profitable and some farmers
consider the benefits of some fungicides sufficient to justify their
routine use. Other farmers could benefit from more extensive use of
fungicides (King, 1977). However, there is a need for much better
information that will allow cost/benefit ratios to be estimated, so
providing a more rational basis for disease control (Cook *et al.*, this
volume). There is probably greater scope for increasing farm profits
by the more efficient use of fungicides than of insecticides and
herbicides. Of pests, only wheat bulb fly regularly decreases yield.
Other pests, of which aphids are perhaps the most important, affect
yield only sporadically. Pests in cereals probably cause a national
yield loss of no more than 2% per annum (C. Collingwood, personal
communication).

Opportunities for improving yield by better weed control are limited
Yield response to broad-leaf weed control is frequently slight (Evans,
1969). A national survey of wild oats (Elliott *et al.*, 1979) has shown
that this weed is now reasonably well controlled and only about a
further 1% of fields are likely to repay chemical treatment. Similarly
herbicides to control black grass are widely used. Future developments
in the use of herbicides are likely to be in their more economical use,
improved selectivity, and in limiting the crop losses that can result
from their misuse.

At present herbicides used to control monocotyledonous weeds represent about a quarter of the total United Kingdom (UK) agrochemical marke
(Steed & Sly, 1979) and farmers spend more than five times as much on
weed control as on disease control. As the greatest gain from pesticid
use in the next few years will probably be from fungicides, there may
be advantages in switching some of the farmers' crop protection
resources from weed control to disease control.

FARM STRATEGIES

The farmer's aim of maximum profit from disease control may be modified
by other considerations:

1.   A desire for minimal decision making.
2.   How easy the control methods are to adopt.
3.   A desire to minimize pesticide use or, at the other extreme,
      to maximize disease control.
4.   Cropping policy and general management factors.

Strategies, which need to suit the individual farm, will be based on:

1.   The farmer's aims.
2.   The control methods available.
3.   How commonly diseases occur.
4.   The expected crop response.

   To choose the best strategy one has to estimate the probable severity
of disease, probable effectiveness of control by the different methods
available and probable economic consequences. To determine such
probabilities, the factors which influence the occurrence and severity
of disease, the effectiveness of control measures, their acceptability
and cost and the likely effect on crop yield and quality, must be
understood.

   The principal methods of disease control are discussed below in the
context of the total farm strategy.

Genetic Resistance

Genetic resistance to disease or tolerance of disease is of considerable
importance. Although it may be argued that pesticides provide a
satisfactory alternative to genetic resistance, they are costly and
their use requires management skills whereas cultivar resistance can be
cheap and is trouble-free to use. It provides a base line to which
other control measures are added. The greater the inherent resistance
of the cultivar, the more effective, and possibly less expensive, other
control measures are likely to be. The imposition by the National
Institute of Agricultural Botany (NIAB) of minimum levels of resistance
to disease in recommended cultivars should help to raise this base line.
Disease may also be decreased if farmers follow advice to diversify the
cultivars they grow, so that large areas are not sown to crops with the
same genetic resistance (Priestley, this volume). Although some
farmers may think it better to sow the cultivar expected to give the
greatest yield and rely on chemicals to control any diseases which
develop (Effland, this volume), this argument makes two assumptions.
The first is that the farmer can predict which cultivar will yield most.
However, trial results consistently demonstrate that cultivar performance is variable, so it can be argued that diversifying cultivars is as
likely to increase average farm yields as it is to decrease them.
The second is that disease can always be satisfactorily controlled by
fungicides. This may not be so because it may not be possible to
achieve adequate control when disease is severe and there may be
practical difficulties in applying pesticides at the optimum time to
the whole of the area needing treatment.

If it is accepted that on many farms there is little to lose and probably something to be gained from sowing more than one cultivar, then the recent development of sowing several cultivars mixed together must also be considered (Wolfe et al., this volume). The cost to the farmer is theoretically small but there may be problems. Cultivars are chosen by the farmer not only for their resistance to disease (and commonly to more than one disease) but also for their yield potential, time of maturity, resistance to lodging or grain shedding, grain quality, resistance to herbicides and suitability to soil and site. Unless cultivars not at present recommended by the NIAB are considered, there are likely to be few cultivar combinations which will be agronomically suitable. Mixtures have so far been tested mainly for the control of barley mildew. Their effectiveness in controlling other diseases of cereals has had much less attention although the work of Jeger et al. (this volume) is encouraging.

There is no evidence to suggest that, in the absence of disease, mixtures of cultivars will yield less than the average of their components grown singly. Therefore routine testing of mixtures in NIAB cultivar trials is unlikely to be necessary. Current evidence is that selection of pathogen races able to attack all components in a mixture (so called "super races") is not an inevitable consequence of their use, particularly if the mixtures are diversified.

Husbandry

Farm practice is often dictated by seasonal events outside the farmer's control and is always subject to the available resources of labour and capital and the expected returns from various cropping systems. The current trend towards growing more winter cereals, which among other things increases autumn work-loads, is justified in farmers' minds by the expectation of better yields, smaller spring work-loads and earlier harvesting. This intensive autumn cereal cropping has led to earlier drilling and reduced cultivations and as a consequence an increased risk of disease. However, farmers are confident that pesticides will enable them to control any diseases which do occur. Problems are also developing with weeds such as barren brome and black grass. It is possible that, in the future, increasing inputs will be required to maintain satisfactory yields of an increased area of winter cereals and that there may then be a return to spring cereals with a need to maximize their yields.

Disease development is affected by crop micro-climate and host susceptibility, both of which can be influenced by husbandry. However, these effects are rarely taken into account when choosing husbandry methods. For example, when deciding how much nitrogen to use and when to apply it, and whether or not to use chlormequat on wheat to decrease the risk of lodging, effects on disease are usually given little consideration. Nevertheless Jenkyn & Finney (this volume) have shown that nitrogen can have a marked effect on disease severity and in the current reassessment of the effects of nitrogen on cereal development and yield, the crops' susceptibility to pathogens should be considered.

Seed Production

Healthy seed is an obvious prerequisite for a healthy crop. Established seed production techniques and testing can ensure that healthy
seed is provided to farmers at a relatively small cost. When home-saved
seed is used, damage by seed-borne diseases is sometimes severe and
farmers should use the seed testing services available before sowing
such seed. In normal circumstances, home-saved seed should be used
only for the first generation after the high quality purchased seed.

Fungicides

A question often asked is whether fungicides should be used prophylactically or curatively. Prophylaxis is honoured by time. For
decades cereal seed in the UK has been treated with mercury without
considering the need for it even though treatment applied only in
response to specific need would use less pesticide and be more
profitable. However, arguments for the two approaches have become too
extreme. If the profit motive is accepted as the basis for using
pesticides then both approaches can be justified according to prevail-
ing circumstances.

Prophylactic treatments are inherently attractive. They simplify
management, they produce a steady demand for pesticides which eases
pesticide production, transport and storage, they avoid the risk that
conditions may prevent treatment even if the need for it is recognized
and their insurance element is psychologically attractive to farmers.
The insurance value is particularly important in crops which have
already incurred high variable costs. Furthermore in crops which have
a high yield potential a small proportional increase in yield will
cover the cost of treatment (Savage, 1977). Broad-spectrum fungicides
or mixtures of fungicides which control a range of pathogens add to the
attractiveness of routine treatment, as do reports of unexplained
yield increases obtained with some broad spectrum sprays even in the
absence of recognized disease.

Where costs are small and methods simple (e.g. mercury seed
treatment) there is little economic argument against prophylactic
treatment and the prospects of extending low cost seed treatment to
control a wider range of diseases is attractive. The economic
arguments for more expensive prophylactic treatments have to be made
on a farm by farm, or probably field by field, basis. Blanket
treatment of all crops in the country is unlikely to be economically
justified except perhaps against barley mildew (Cook *et al*., this
volume). It is also undesirable because it risks damage to the environment and increases the risk of pathogen resistance and a consequent
decrease in the effective life of fungicides (Conway & Comins, 1979).
By using them either sequentially or in mixtures the development
of resistance may be delayed (Dekker, this volume). Even if prophylactic treatments are used there are clearly advantages in not applying
them to crops in which disease is unlikely to develop. Jenkins (1977)
and Jordan & Tarr (this volume) have discussed criteria, for

mildew and eyespot respectively, which are useful in deciding whether prophylactic treatment can be justified in particular fields.

Deciding when to use curative treatments is difficult because assessing the need for treatment often requires frequent crop inspection. The criteria for treatment must be readily understood by the farmer or his adviser. Applying sprays at the correct time to large areas may be difficult although advances in spray technology may help. Even curative treatments may be applied needlessly if conditions after treatment no longer favour the pathogen. Nevertheless, using all available information, it should be possible to base treatment on specific need and so more nearly restrict fungicide use to crops most likely to respond economically and thus counter the accusation of over-use of pesticides.

CONCLUSIONS

Pests, diseases and weeds need not be serious constraints on cereal yields but there is scope for improving fungicide use.

Strategies for disease control on the farm should be based on healthy seed of genetically resistant cultivars, supported by cultivar diversification. Whilst diseases can be much influenced by crop husbandry, there are often other considerations which argue against fully adopting techniques that give minimum disease.

Current farm strategies rely heavily on fungicides, and their use should be diversified as much as possible. Whilst prophylactic treatment can be justified for some diseases on some fields, better criteria are needed so that less fungicide is used unnecessarily. Criteria for curative treatments should be simple to understand and easy to determine. Advances in spray technology are required to allow the maximum number of crops to be sprayed at the optimum time.

REFERENCES

Alcock M.B. (1967) Understanding crop yields. *Arable Farmer* 1, 42-5.
Austin R.B. (1978) Actual and potential yields of wheat and barley in the United Kingdom. *ADAS Quarterly Review* 29, 76-87.
Conway R.R. & Comins H.N. (1979) Resistance to pesticides. Two lessons in strategy from methematical models. *Span* 22, 53-5.
Elliott J.G., Church B.M., Harvey J.J., Holroyd J., Hulls R.H. & Waterson H.A. (1979) Survey of the presence and methods of control of wild oat, black grass and couch grass in cereal crops in the United Kingdom during 1977. *Journal of Agricultural Science, Cambridge,* 92, 617-34.
Evans S.A. (1969) Spraying of cereals for the control of weeds. *Experimental Husbandry* 18, 102-9.
Evans S.A. (1977) The place of fertilizers in "blueprints" for the production of potatoes and cereals. *Fertilizer Use and Production of*

Carbohydrates and Lipids, Proceedings of the 13th Colloquium of the International Potash Institute, York, United Kingdom, pp.231-41.

Fiddian W.E.H. (1977) Factors affecting cereal yield. *Proceedings 1977 British Crop Protection Conference - Pests and Diseases* 3, 755-61.

Jenkins J.E.E. (1977) Future strategy for pest and disease control in cereal production systems. *Proceedings 1977 British Crop Protection Conference - Pests and Diseases* 3, 785-93.

Jenkins J.E.E. (1978) Current work: Existing comprehensive and multi-disciplinary studies (ADAS). *Maximising Yields of Crops (Proceedings of a symposium)* Her Majesty's Stationery Office, London, pp. 214-5.

King J.E. (1977) The incidence and economic significance of diseases in cereals in England and Wales. *Proceedings 1977 British Crop Protection Conference - Pests and Diseases* 3, 677-87.

Savage M.J. (1977) Prevention or cure? The case for or against prophylactic treatment in crop protection. *Proceedings 1977 British Crop Protection Conference - Pests and Diseases* 3, 672-5.

Steed J.M. & Sly J.M.A. (1979) Pesticide Usage. Preliminary Report Arable Crops 1977. Ministry of Agriculture, Fisheries & Food, Harpenden Laboratory. 33 pp.

Disease control on the farm

T. D. G. PEARSON

Cockrup Farm, Coln St. Aldwyns, Cirencester, Gloucestershire

INTRODUCTION

My object as the manager of a commercial farm is to make money but
because long-term business planning is difficult unless there is some
consistency in returns, I aim, in practice, for maximum *repeatable*
profitability. However, with many inputs, it is difficult to critically
evaluate the economics of using them. Many obey the law of diminishing
returns but the economic optimum may vary from year to year and be
difficult to predict. Other inputs are "lumpy" and not normally used in
small increments so one is faced with a simple question as to whether or
not they should be used. The decision to use them is obviously based on
the expected economic return but is also commonly influenced by their
insurance value and the sense of security which results. Many of the
measures used to control diseases fall into this category but it must be
remembered that these represent a relatively small proportion of the
many factors involved in growing a crop or managing a farm. This
chapter is confined to my own limited experience with cereals, predominantly winter wheat and winter barley, and aims to show the relevance
of disease control in producing maximum repeatable profits.

FACTORS AFFECTING PROFITABILITY

Profitability is determined by the value of the output and the cost of
the inputs, both of which are influenced by a variety of factors. Many
of them are under the grower's control, including choice of cultivar
(the yield potential and quality of which will largely determine end-
product value), chemicals and fertilizers, and also many of the fixed
costs, although the latter are determined by long term decisions and not
easily changed in the short term. Other factors are less under his
control and can be regarded as risks. These include the weather,
cultivar susceptibility and the possible breakdown of resistance to
disease, effects of rotation, unknown phytotoxicity of chemicals, and
even the ultimate price paid for the product. There may also be
interactions between factors, perhaps unknown, which may either
increase or decrease yield.

Jenkyn J. F. & Plumb R. T. (1981) *Strategies for the Control of Cereal Disease*

DISEASES AND THEIR CONTROL

An important influence on yield, which may interact with many other
factors and hence influence many decisions a grower must make, is
disease. Those diseases of most significance in cereals, with the
notable exeption of barley yellow dwarf virus, are caused by fungi.
Control relies almost entirely on host resistance, husbandry and the use
of chemicals (Table 1); biological control, although attractive in
theory, is the least practicable. By sensibly integrating all the
available control measures, diseases are now largely under management
control; the likely effects of disease can be assessed and decisions to
control them based on the expected economic return. It is not always
possible to make the right decisions but with increasing knowledge the
task becomes much easier.

Table 1. Principal factors, within the grower's control, which
influence disease incidence and severity.

1. Choice of cultivar	(a)	Resistance in specific cultivars
	(b)	Diversification
	(c)	Mixtures
2. Husbandry	(a)	Stubble hygiene
	(b)	Density of sowing
	(c)	Date of sowing
	(d)	Rotation
	(e)	Fertilizer use
3. Use of chemicals	(a)	In response to specific need ("Fire brigade" approach)
	(b)	Routinely

Choice of cultivars

Accurate information on cultivar susceptibility to disease is readily
available from the National Institute of Agricultural Botany (NIAB).
They also test other agronomic characters, including yield and quality,
so that the grower can easily recognise the strengths and weaknesses of
different cultivars. However, it is important to note that many farmers
do find that it can be more profitable to grow some cultivars that are
not recommended by NIAB because of extreme susceptibility to particular
diseases, provided that they use appropriate crop protection measures.
Commonly the extra effort and expense is well rewarded. Many diseases
are favoured by particular climatic conditions but by following advice
given by the Agricultural Development and Advisory Service (ADAS) and
NIAB it is possible to choose cultivars most suited to particular areas.
NIAB also give specific advice on cultivar diversification (Priestly,

this volume), intended to lessen the risks of severe epidemics. The use of mixtures (Wolfe *et al.*, this volume) may also help to decrease diseases, and the losses they cause, but my fear is that this could be open to abuse and used for covering up cultivar impurity.

Husbandry

Cultural practices can have a marked influence on pathogens, especially on their survival. One seldom adopts a practice solely to minimise disease but these effects are important because they may have repercussions on the total disease-control strategy. For example, stubble hygiene is very important but the choice of whether to burn, spray, minimally cultivate or plough, or a combination of these, is also influenced by the costs of these operations.

Sowing early in autumn has long been recognised as likely to increase the risk of infection by diseases because, at least in part, there is very little break between successive cereal crops. However, in the Cotswolds, where I farm, early sowing is considered essential if we are to fully exploit the high yield potential of winter wheat and winter barley. In these circumstances the judicious use of chemicals provides a means of overcoming the problems created not only by early sowing but also by other cultural practices such as high plant densities and liberal use of nitrogen (both of which increase the risk of damage by disease), and thus realise the potentially higher returns of intensive systems.

Where cereals are grown frequently, adopting the correct disease control strategy is especially important. For example, eyespot is a particular problem for continuous cereal growers like me. However, effective fungicides have enabled us to overcome the limitations to yield caused by this pathogen, while a better understanding of take-all and its decline during a run of cereals has allowed the grower to avoid the worst effects of this disease, at little extra cost.

It has been known for some time that increased use of nitrogen fertilizer encourages several fungal diseases, and interactions between nitrogen use and fungicides are not uncommon (Jenkyn & Finney, this volume). The sensible use of fungicides on certain new high yielding cultivars (e.g. Hobbit) improves their response to applied nitrogen and increases the amount which it is economic to apply.

Use of chemicals

Although the environment must be protected and I favour methods of pest and disease control which have the least deleterious effects, it is difficult to avoid the use of fungicides in the face of ever increasing economic pressures. Admittedly fungicides can be very costly but, used correctly, they can also be very profitable. The decision to use them can have implications beyond their immediate effect on disease and the consequences of this use must be carefully evaluated. Unfortunately much of the information on which to base practical fungicide use in the

United Kingdom is still lacking. The problem is aggravated because
many growers try to run before they can walk, and are also influenced
by accounts of the benefits of disease control on the continent. It
also seems to me that some governmental bodies are not always fully
aware of current farming practice so that their results are less useful
than they might otherwise be. For example, although autumn control of
mildew on winter barley may not always be justified when the crop is
sown in mid-October, our experience shows that it is essential when
crops are sown in September as is now normal in the Cotswolds. Yet
trials are still being sown in mid-October. In my own district, farmers
have now formed a cereal centre to obtain locally relevant information
on the costs and benefits of disease control and the profitability of
different cereal-growing systems.

Improved methods for forecasting diseases are undoubtedly needed.
Nevertheless, even with present limited knowledge, it is possible to
develop disease control programmes which make the best use of all
available methods and give an economic return while still providing
reasonable insurance against unexpected losses. The broad spectrum of
activity of some chemicals, which allows combinations of diseases to
be controlled at no extra cost, is attractive to many growers.
Similarly, there is a great temptation to apply mixtures of chemicals
but in the absence of accurate information there is a risk that growers
will use mixtures which are incompatible, either physically or biolog-
ically. More information on suitable mixtures is desperately needed.

Ideally crops, and especially those most at risk from disease, should
be examined as frequently as possible. However, such inspection is not
always possible and irreparable damage may have been caused by the time
diseases are noticed. Better forecasting methods might decrease the
need for such frequent crop inspection and would be expecially useful
for pathogens, such as the rusts, which are capable of multiplying very
rapidly and causing severe damage. Even with accurate disease forecasts, if weather conditions are unsuitable it may still be impossible
to apply fungicides when they are needed.

CONTROL OF WHEAT DISEASES ON COCKRUP FARM IN 1979

To illustrate our approach to cereal disease control I will briefly
outline what we do, using 1979 cropping data. We grow c. 243 ha of
winter wheat but at present we grow no quality milling wheat because
the existing premium on it is insufficient to compensate for its smaller
yield. Wheat sown after a break (c. 25% of the total) was of the
cultivar Mardler. These crops are at low risk from eyespot and are only
sprayed with CCC and a carbendazim fungicide if crop inspection shows
severe eyespot to be present. Mildew is not a major problem on this
cultivar but maneb plus carbendazim is applied routinely at about growth
stage (GS) 50 (Zadoks et al., 1974), primarily to control septoria and
sooty moulds although it will give some control of both mildew and
rusts. Crops following wheat in 1978 (25%) were of the cultivar
Flanders. These were sprayed routinely with CCC and carbendazim at

GS 30-31 to control eyespot and with maneb plus carbendazim at about GS 50 to control septoria and sooty moulds. The remaining area (*c.* 120 ha) is continuous wheat, mostly sown to Hobbit, a cultivar susceptible to mildew. These crops are sprayed routinely with CCC and carbendazim at GS 30-31, to control eyespot, and with triadimefon at around GS 37 if mildew or yellow rust are seen. However, triadimefon plus captafol are applied routinely at GS 50 to control mildew, septoria and the sooty moulds.

In my experience, resistance to septoria is not as good as NIAB ratings suggest and this is why we apply sprays to control this disease as a routine. Yellow and brown rust are only occasionally a problem and sprays to control them are only used when crop monitoring shows them to be necessary.

At 1979 prices, the total cost of chemicals, excluding cost of application, was £12.50, £21.25 and £36.25 per ha for first, second and continuous wheats respectively, which demonstrates that effective disease control is possible without excessive reliance on chemicals. Our results with Hobbit have shown how the combined use of extra inputs, especially fungicides and fertilizers, allow one to better exploit the much greater yield potential of some modern cultivars.

CONCLUSIONS

The cereal grower faces many difficult decisions and how best to cope with disease is only one. Therefore it is not surprising that many can see management advantages in adopting an insurance approach to disease control at the risk of using excessive amounts of chemical. Hopefully, with some co-operation between the grower and the scientist a more critical approach to the use of chemicals will develop, so decreasing the risks of pathogen resistance, while maintaining the growers' returns.

REFERENCES

Zadoks J.C., Chang T.T. & Konzak C.F. (1974). A decimal code for the growth stages of cereals. *Weed Research* 14, 415-421.

Index